ROBERT C. WILLIGES

THE BIOMECHANICAL BASIS OF ERGONOMICS

THE BIOMECHANICAL BASIS OF ERGONOMICS
Anatomy Applied to the Design of Work Situations

E. R. TICHAUER

A WILEY-INTERSCIENCE PUBLICATION
John Wiley & Sons, New York • Chichester • Brisbane • Toronto

Copyright © 1978 by John Wiley & Sons, Inc.

All rights reserved. Published simultaneously in Canada.

Reproduction or translation of any part of this work beyond that permitted by Sections 107 or 108 of the 1976 United States Copyright Act without the permission of the copyright owner is unlawful. Requests for permission or further information should be addressed to the Permissions Department, John Wiley & Sons, Inc.

This publication is designed to provide accurate and authoritative information in regard to the subject matter covered. It is sold with the understanding that the publisher is not engaged in rendering legal, accounting, or other professional service. If legal advice or other expert assistance is required, the services of a competent professional person should be sought.

From a Declaration of Principles jointly adopted by a Committee of the American Bar Association and a Committee of Publishers.

Library of Congress Cataloging in Publication Data:

Tichauer, E R
   The biomechanical basis of ergonomics.

   "A Wiley-Interscience publication."
   Bibliography: p.
   Includes index.
   1. Human engineering. 2. Biomechanics.
I. Title.

TA166.T5   620.8′2   77-28807
ISBN 0-471-03644-7

Printed in the United States of America

10 9 8 7 6 5 4 3 2 1

Preface

Ergonomics, the discipline dealing with the interaction—physical as well as behavioral—between man, his workplace, his tools, and the general environment, is a very broad field. It utilizes, as tributaries, so many aspects of the biological, behavioral, medical, and technological sciences that complete coverage of the entire field is difficult in a book of manageable size. Furthermore, the term "ergonomics" has different meanings in different environments. In some situations it is almost synonymous with "human factors" or "engineering psychology." In others the practice of the discipline is closely akin to the application of work physiology. Likewise, specialists from numerous disciplines are practitioners in ergonomics. These include occupational physicians, industrial engineers, safety specialists, industrial hygienists, industrial designers, rehabilitation professionals, and many others. The common denominator of the ergonomic know-how of all these professionals is usually some training in physiology and the behavioral sciences. The anatomy of function, the structural basis of human performance—also known as biomechanics—is less universally taught. Nevertheless, occupational disease, accidents, and low levels of productivity are more often than not the result of inadvertent neglect of simple biomechanical principles in the design of equipment or workplace layout.

Only a small number of specialists are dedicated exclusively to research and practice of occupational biomechanics. However, numerous practitioners of other disciplines concerned with the health, welfare, and performance of the working population require some understanding of the principles of functional anatomy applied to the design and improvement of tasks common in manufacturing and service industries. This treatise, while not exhaustive, deals with the majority of those cases in which the professional not specialized in ergonomics has to make immediate decisions and develop ad hoc solutions to ergonomic problems without recourse to lengthy experimentation or the advice of a specialist.

The analysis of problems created by the general industrial environment, including those stemming from noise, light, and climate, has been omitted. Specialized reference works on these are available. The subject matter in this book was selected on the basis of questions asked frequently by practitioners in industry and those problems encountered most frequently in the course of my own professional practice or discussions with my students.

It is hoped that the application of the material presented here will be of assistance to practitioners in industry endeavoring to ensure that the people under their care perform efficiently and economically while enjoying, at the same time, the highest possible levels of health, physical well-being, and safety.

E. R. Tichauer

New York, New York
April 1978

Contents

1 Historical Background   1
     1.1 Ergonomic Stress Vectors, 3
     1.2 Definition of "Ergonomist", 4

2 The Anatomy of Function   4
     2.1 Anatomical Lever Systems, 5
     2.2 Occupational Kinesiology, 9
     2.3 Application of Kinesiology to Workplace Layout, 10
     2.4 Optimal Placement of Equipment Controls, 11

3 Physiological Measurements   16
     3.1 Metabolic and Quasi-Metabolic Measurements, 17
     3.2 Electromyographic Work Measurement, 19
     3.3 Electromyographic Technique, 22
     3.4 Interpretation of Myograms, 27

4 Work Tolerance   31
     4.1 The Prerequisites of Biomechanical Work Tolerance, 33
     4.2 The Postural Correlates of Work Tolerance, 33
     4.3 The Engineering of the Man-Equipment Interface, 37
     4.4 The Development of Effective Kinesiology, 43

5 Manual Materials Handling and Lifting   47
     5.1 Elemental Analysis of Lifting Tasks, 48
     5.2 Queueing Situations, 58

6 Hand Tools   59

    6.1 Basic Considerations in Tool Evaluation, 59
    6.2 The Anatomy of Function of Forearm and Hand, 62
    6.3 Elemental Analysis of Hand Movements, 67
    6.4 Trigger-Operated Tools, 69
    6.5 Miscellaneous Considerations, 70

7 Chairs and Sitting Posture   71

    7.1 Anatomical, Anthropometric, and Biomechanical Considerations, 71
    7.2 Adjustment of Chairs on the Shop Floor, 76
    7.3 Ancillary Considerations, 77

8 Ergonomic Evaluation of Work Situations   78

    8.1 Historical Evaluation, 79
    8.2 Analytical Evaluation, 81
    8.3 Projective Evaluation, 82

Acknowledgments   85
Glossary   86
References   93
Index   97

THE BIOMECHANICAL BASIS OF ERGONOMICS

1 HISTORICAL BACKGROUND

The systematic study of the ill effects on man of poorly designed work situations is by no means of recent origin. Ramazzini (1), around 1700, elaborated on the disastrous effects of work stress and its disabling consequences for those engaged in physical labor. He wrote: "Manifold is the harvest of diseases reaped by craftsmen. . . . As the . . . cause I assign certain violent and irregular motions and unnatural postures . . . by which . . . the natural structure of the living machine is so impaired that serious diseases gradually develop." At that time, however, and for centuries to come, there existed no technique for the study of the anatomy of function of the living body. Also laborers were considered to be expendable, and occupational disease was ordinarily rewarded by dismissal.

Thus with few noteworthy exceptions, such as Thackrah (2), industrialists, physicians, and even the precursors of today's social scientists, remained insensitive to the exposure of the workforce to ergogenic disease produced by mechanical noxae until World War I, when labor became a scarce resource essential to the very survival of the warring nations. This stimulated physiologists (3), as well as psychologists, to embark on an intensive study of the effects of working conditions on human performance and well-being.

Experimental methods for these disciplines were already well established, and their application to problems of man at work was quite easy. One of the man-oriented life sciences lagged sadly behind, however: anatomy, which had remained a cadaver-based, geographic discipline. Thus physiological responses and behavioral reactions to the demand of work situations were investigated for several decades before the exploration of the structural and mechanical basis of musculoskeletal performance.

Change was pioneered by Adrian (4) who recorded electromyograms during movement, relating them to kinesiological events. The Great Depression slowed down progress in these avenues of scientific inquiry, but during World War II biological and behavioral scientists again had the opportunity to lead the quest for an improved utilization of human resources.

During the postwar period work physiology, industrial psychology, and other specialties consolidated into a broad discipline dedicated to the study of man at work: ergonomics. The Ergonomics Research Society was organized in 1949 to serve the need of those professionals in a variety of disciplines concerned with the effects of work on man.

The founders of the new society argued about the most suitable name, suggesting (5) "Society for Human Ecology" and "Society for the Study of Human Environment."

Finally, "ergonomics" was adopted as a term neutral with respect to the relative importance of the behavioral sciences, physiology, and anatomy. An anatomical methodology for the study of work did not yet exist. Thus the thrust of ergonomics had to stem initially from the other two disciplines. However a speedy development of experimental biomechanics and novel instrumentation available added "live body anatomy" to the spectrum of research resources available to ergonomists.

Gradually, as interest grew in the structural basis of human performance, the electrogoniometer was perfected (6). This made it possible to use simultaneously goniometry and Adrian's myography when investigating the relationships between muscular activity and ensuing movements. Thus information about the functional anatomy of the living body could be applied by ergonomists to the problems germane to both occupational health and industrial productivity. One of the numerous pioneers was Lundervold (7), who related myoelectric signals obtained from typists to posture and hand usage (this was perhaps the first comprehensive biomechanical analysis of a common work situation). Thus electrophysiological kinesiology was developed.

Utilizing procedures of electrophysiological kinesiology, Tichauer (8) added the biomechanical profile to the techniques available for the study of interaction between

Figure 1 A worker surrounded by external physiological and mechanical environment, which must be matched to his internal physiological and biomechanical environments, symbolizes the concept of modern ergonomics (biomechanics). From Reference 79.

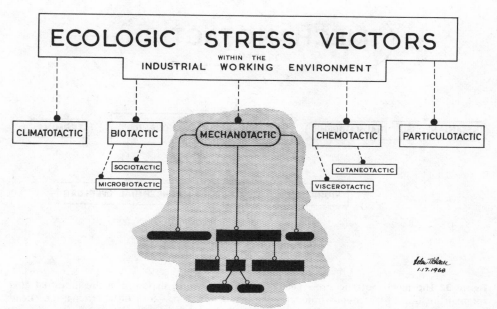

Figure 2 The scheme of ecologic stress vectors common to all working environments. Work stress is derived from contact with climate, contact with living organisms such as fellow man or microbe, contact with chemical elements or compounds, contact with hostile particles such as silica, or asbestos and finally, contact with mechanical devices. From Reference 13.

worker and industrial environment. Meanwhile, behavioral scientists such as Lukiesh and Moss (9) pioneered research into the effects of light and illumination on human performance. Work physiologists like Belding and Hatch (10) described and explained the effects of climate on working efficiency and studied noise and its effects on workers, not only with respect to deafness, but also in relation to many other physiological parameters (11). The study of performance decrements due to adverse working conditions became the object of an entire new school of students of human fatigue (12).

Thus today, knowledge gathered from many tributary sciences has been blended into a unified discipline dealing with the effects of work on man.

1.1 Ergonomic Stress Vectors

The basic philosophy of ergonomics (Figure 1) considers man to be an organism subject to two different sets of laws: the laws of Newtonian mechanics, and the biological laws of life. It is part of this philosophy to postulate that man in work situations is surrounded by the external physical working environment, and inside the human body the "internal biomechanical environment," an array of levers and springs also known as the musculoskeletal system, responds to the demands of the task. The stress vectors commonly acting upon man in the industrial environment have been set out in flow diagram form (Figure 2 and 3).

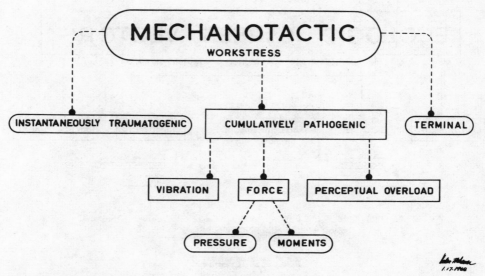

Figure 3 The mechanotactic stress vectors leading to hazard exposure in the industrial environment include instantaneous traumatogenesis (e.g., an arm is torn off); terminal (i.e., death occurs immediately) and, most frequently, cumulative pathogenesis. The latter term describes the gradual development of disability or disease through repeated exposure to mechanical stress vectors over extended periods of time. From Reference 13.

1.2 Definition of "Ergonomist"

An ergonomist is a professional trained in the health, behavioral, and technological sciences and competent to apply them within the industrial environment for the purpose of reducing stress vectors sufficiently to prevent the ensuing work strain from rising to pathological levels or producing such undesirable by-products as fatigue, careless workmanship, and high labor turnover. Ergonomics as a discipline aims to help the individual members of the workforce to produce at levels economically acceptable to the employer while enjoying, at the same time, a high standard of physiological and emotional well-being.

2 THE ANATOMY OF FUNCTION

Anatomy is concerned with the description and classification of biological structures. Systematic anatomy describes the physical arrangement of the various physiological systems (e.g., anatomy of the cardiovascular system); topographic anatomy describes the arrangement of the various organs, muscular, bony, and neural features with respect to each other (e.g., anatomy of the abdominal cavity); and functional anatomy focuses on the structural basis of biological functions (e.g., the description of the heart valves and ancillary operating structures; the description of the anatomy of joints). As distinct and different from the aforementioned categories, the anatomy of function is concerned with the analysis of the operating characteristics of anatomical structures and systems when these interact with physical features of the environment, as is the case in the performance of an industrial task. Whenever the "motions and reactions inventory" demanded by the external environment is not compatible with the one available from the internal biomechanical environment, discomfort, trauma, and inefficiency may arise (3).

The anatomy of function is the structural basis of human performance, thus it provides much of the rationale by which the output measurements derived from work physiology and engineering psychology can be explained.

2.1 Anatomical Lever Systems

The neuromuscular system is, in effect, an array of bony levers connected by joints and actuated by muscles that are stimulated by nerves. Muscles act like lineal springs. The velocity of muscular contraction varies inversely as the tension within the muscles. With very few exceptions, lever classifications and taxonomy in both anatomy and applied mechanics are identical. Each class of anatomical levers is specifically suited to perform certain types of movement and postural adjustment efficiently, with undue risk of accidents or injury but may be less suited to perform others. Therefore a good working knowledge of location, function, and limitation of anatomical levers involved in specific occupational maneuvers is a prerequisite for the ergonomic analysis and evaluation of most man–task systems.

First-class levers have force and load located on either side of the fulcrum acting in the same direction but opposed to any force supporting the fulcrum (Figure 4). This is exemplified by the arrangement of musculoskeletal structures involved in head movement when looking up and down. Then the atlantooccipital joint acts as a fulcrum of a first-class lever. The muscles of the neck provide the force necessary to extend the head. This is counteracted by gravity acting on the center of mass of the head, which is located on the other side of the joint, hence constitutes the opposing flexing weight.

First-class levers are found often where fine positional adjustments take place. When standing or holding a bulky load, static head movement in the mid-sagittal plane produces the fine adjustment of the position of the center of mass of the whole body, necessary to maintain upright posture. Individuals suffering from impaired head movement (e.g., arthritis of the neck), should not be exposed to tasks in which inability to maintain postural equilibrium constitutes a potential hazard. Likewise, workplaces where unrestricted head movement is difficult should be provided with chairs or other means of postural stabilization. Special attention should be paid to any feature within the working environment that may cause head fixation (e.g., glaring lights). Sometimes even an unexpected acoustic stimulus, such as a friendly "hello" directed at an individual carrying a heavy and bulky load, may cause inadvertent sideways movement of the head, which can interfere with postural integrity and result in a fall.

Second-class levers have the fulcrum located at one end and the force acting at the

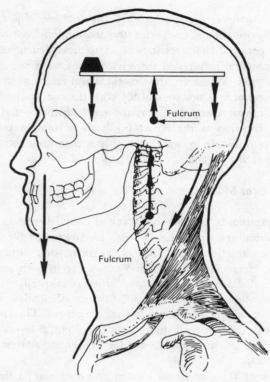

Figure 4 The action of the muscles of the neck against the weight of the head is an example of a first-class lever formed by anatomical structures. The atlantooccipital joint acts as a fulcrum. Adapted from Reference 48.

other end, but in the same direction as the supporting part of the fulcrum. The weight acts on any point between fulcrum and force in a direction opposed to both of them. Second-class levers are optimally associated with ballistic movements requiring some force and resulting in modifications of stance, posture, or limb configuration. The muscles inserted into the heel by way of the Achilles tendon (i.e., force) and the weight of the body transmitted through the ankle joint and the weight of the big toe (i.e., fulcrum) are a good example of a second-class lever system used in locomotion (Figure 5). The movements of this type of lever are never very precise, therefore foot pedals should have adequately large surfaces and their movement should be terminated by a positive stop rather than by relying on voluntary muscular control. Another example of a second-class lever is provided by the structural arrangement of the shoulder joint. Here the head of the humerus acts as a fulcrum, the anterior and posterior heads of the deltoid provide the force; the "weight" is provided by the inertia of the mass of the arm. Hence it follows that a shoulder swing moves the hand to a rather indeterminate location; thus such a movement should be terminated by a positive stop or, alternatively, when an object has to be dropped at the end of a motion, the receptacle should be sufficiently large, or have a flared inlet, to overcome the kinesiological deficiency.

Third-class levers have the fulcrum at one end, and the weight acts on the other end in the same direction as the supporting force of the fulcrum. The "force" acts on any point between weight and fulcrum, but in a direction opposed to them both. Tasks that require the application of strong but voluntarily graded force are best performed by this type of anatomical lever system. Holding a load with forearm and hand when the brachialis muscle acts on the ulna, with the elbow joint constituting a pivot, is a typical example (Figure 6).

Torsional levers are a specialized case of third-class lever. Here the axis of rotation of a limb or long bone constitutes a fulcrum. The force generating muscle of the system is inserted into a bony prominence and produces rotation of the limb whenever the muscle contracts. The "weight" is represented by the inertia of the limb plus any external torque opposing rotation. An example is the supination of the flexed forearm (Figure 7). Here the fulcrum is the longitudinal axis of the radius; the force is exerted by the biceps

Figure 5 The ankle joint, as an example of an anatomical second-class lever system. The fulcrum is located at the base of the big toe. Adapted from Reference 48.

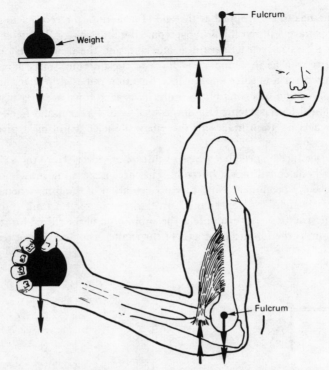

Figure 6 An anatomical third-class lever is formed between ulna and humerus. The brachialis muscle provides the activating force, the fulcrum is formed by the trochlea of the humerus. Adapted from Reference 48.

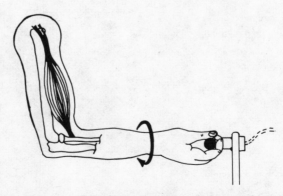

Figure 7 The torsional lever system involving radiohumeral joint, biceps, and a resistance against supination of the forearm can be employed with advantage in such operations as closing of valves.

muscle inserted into the bicipital tuberosity of the radius. The opposing load is made up by the inertia of the forearm and hand, plus the resistance of, for example, a screw being driven home. Tasks to be performed for strength and precision and at variable rates of speed are best assigned to torsional lever systems.

For identification and classification of other lever systems involved in specific occupational maneuvers, standard text books on kinesiology (14–16) should be consulted.

2.2 Occupational Kinesiology

Occupational kinesiology is the discipline concerned with the basic study of human movement and its limitations in work situations. Unfortunately, with the exception of brief monographs (17), all texts and reference books in the field of kinesiology relate either to athletics or rehabilitation medicine.

Kinesiology describes the laws and quantitative relationships essential for the understanding of the mechanisms involved in human performance, either of individuals or of groups of individuals interacting with one another (i.e., a working population). Its basic tributaries are anatomy, physiology, and Newtonian mechanics. It describes and explains the behavior of the whole body, its segments, or individual anatomical structures in response to intrinsic or extrinsic forces.

The student of kinesiology should be thoroughly familiar with the nomenclature and organization of mechanics. Here statics is concerned with the generation and maintenance of equilibrium of bodies and particles. In the context of kinesiology, "bodies" are generally synonymous with anatomical structures and "particles" become anatomical reference points. Likewise, the biodynamic aspects of kinesiology are explained through kinematics, which is concerned with the geometry and patterns of movements, but not with causative forces producing motion. Kinetics, on the other hand, deals with the relation between vectors and forces producing motion and also with the output from body segments in terms of force, work, and power, including the resulting changes in temporal and spatial coordinates of anatomical reference points.

Before the kinematics basic to a specific kinesiological maneuver can be explored, the kinematic element involved must be identified. A kinematic element consists of bones, fibrous and ligamentous structures pertaining to a single joint inasmuch as they affect the geometry of motion. Because kinematics is not concerned with forces, muscles do not normally form part of a kinematic element. As an example, consider the kinematic element of forearm flexion, which consists of the humerus, the ulna, the joint capsule of the humeroulnar joint, and associated ligaments.

Kinematic elements can have several degrees of freedom of motion; the higher this number is, the greater the variety of movements that can be produced. However accurate movements produced by elements possessing a high degree of freedom require a proportionally higher level of skill: for example, it is easier to position the hand accurately by means of humeroulnar flexion than by a shoulder swing. Likewise, the higher the degree of freedom, the greater the influence of musculoskeletal configuration on the effectiveness of movement. As an example, the following are mentioned:

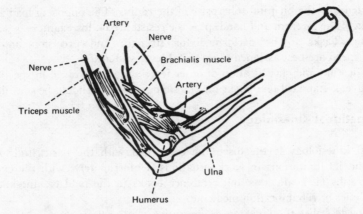

Figure 8 The kinetic element of forearm flexion and extension. From Reference 48.

1. Humeroulnar joint: 1 degree of freedom (i.e., flexion); effectiveness quite independent of general musculoskeletal configuration.
2. Whole elbow joint: 2 degrees of freedom (i.e., flexion, pro/supination); effectiveness somewhat dependent on musculoskeletal configuration.
3. Hip joint: 3 degrees of freedom (i.e., flexion, ad/abduction, circumduction); the effectiveness of this chain actually, in a number of situations, such as walking, is quite dependent on musculoskeletal configuration.

Often in workplace layout, kinematic elements are considered in the initial planning of the geometry of the work situation; but when activity tolerance and other work and effort relationships are important, kinetic elements must be considered. These include constituents of kinematic elements, but in addition they incorporate muscular structures as well as the stimulating nerves and nutrient blood vessels because these affect the immediacy, strength, and endurance aspects of a specific kinesiological maneuver (Figure 8).

2.3 Application of Kinesiology to Workplace Layout

Kinesiological concepts can be applied with advantage to the design of work situations when it is essential to minimize physical stress and fatigue. When analyzing motion patterns incidental to the performance of industrial tasks, it is desirable to start with the preparation of a length–tension diagram (Figure 9). This is produced by making the protagonist muscle of the kinetic element performing the motion contract isometrically against measured resistance at different points of the motion's pathway. We plot on the x-axis the included angle between the major bony elements involved. On the y-axis the maximal force exerted during contraction in each position is recorded. Generally, the only range of joint movement to be utilized in workplace layout is that which coincides

ERGONOMICS

with the highest portion of this curve. The length–tension diagram will show that most kinetic chains can be utilized effectively only throughout a very narrow angle of joint movement, and it will identify the limits of this range.

The force–velocity diagram is another useful graphic representation of the effectiveness of joint movement (Figure 10). The rate of change in joint configuration is plotted against maximum forces developed by the muscle, forming roughly a negative exponential curve. The zero velocity value corresponds to the maximum of force. At maximum velocity the force exerted by a muscle approaches zero as a limit. Hence high velocity and high muscular forces are mutually exclusive. The plotting of a force–velocity curve is a complex undertaking not always feasible under field conditions or even in the laboratory. However an awareness of its general configuration often serves to protect the practicing ergonomist against the selection of ineffective musculoskeletal configurations in workplace layout.

2.4 Optimal Placement of Equipment Controls

In the operation of many equipment controls and other industrially used devices there is no noticeable displacement of anatomical reference points while muscles contract. Under such conditions, isometricity of movement may be assumed. This implies that forces exerted by protagonist and antagonist muscles acting on a limb are in equilibrium with each other, even when exerting maximal force. Occasionally—for example in the raising

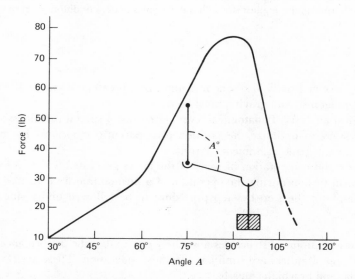

Figure 9 Length–tension diagram produced by flexion of the forearm in pronation. "Angle" refers to included angle between the longitudinal axes of forearm and upper arm. The highest parts of the curve indicate the configurations where the biomechanical lever system is most effective.

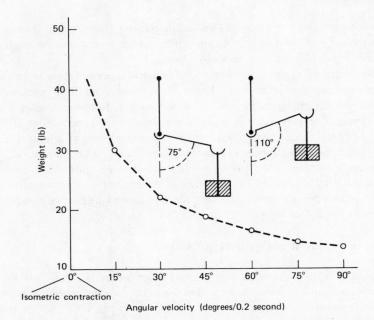

Figure 10 Force-velocity curve of elbow flexion; forearm in supination. Excursion of the limb through the narrow range between 75 and 110° of included angle between forearm and upper arm. The diagram shows that high strength and high velocity of movement are mutually exclusive conditions. Furthermore, the highest strength is developed under conditions of zero velocity (i.e., isometricity).

of a leg—the force of gravity also acts on a joint; in this case, the sum of all three forces, protagonist, antagonist, and gravity, must be zero.

Kinesiological analysis of anatomical lever systems is of special usefulness whenever it becomes necessary to optimize the position of apparently innocuous but nevertheless potentially traumatogenic equipment controls.

The analysis and computation of some of the forces generated within the kinetic element of forearm flexion during the operation of a pushbutton are described. This may serve as example of the step by step procedure to be followed under similar circumstances.

Step 1. Those anatomical structures absolutely essential to the performance of the task are identified and all others are eliminated from consideration. These are (Figure 8) the humerus, ulna, and brachialis muscle.

Step 2. Those forces that, if excessive, may lead to anatomical failure are identified. In this example these are tension in the brachialis muscle and thrust on the elbow joint.

ERGONOMICS

Step 3. A force diagram (such as is used in mechanics) is drawn true to scale (Figure 11).

Step 4. The mechanical assumptions essential to the solution of the problem are made. As the condition of isometricity exists, the sum of torques acting clockwise on the elbow joint must be equal to the sum of those acting counterclockwise. (The symbols used in the following computations are defined in the legend to Figure 11).

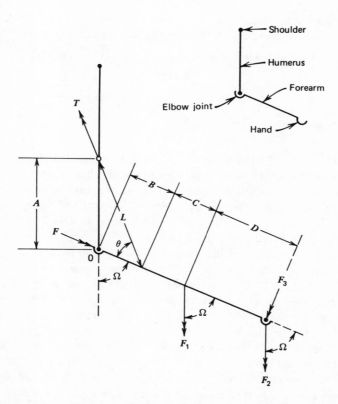

Figure 11 Vector diagram used in the computation of forces acting on elbow joint and tension generated in brachialis muscle when operating a push button. A, distance from the average origin of the brachialis to the center of rotation of the humero-ulnar joint; B, distance from the center of rotation to the average insertion of the brachialis on the ulna; C, distance from the insertion to the center of gravity of the forearm; D, distance from the center of gravity to the application of the load (either F_2 or F_3); F_1, weight of the forearm and hand; F_2, a weight in the hand; F_3, a force normal to the hand at all angles; T, tension in the brachialis; F, compressive force exerted on the elbow joint; L, distance between origin and insertion of brachialis; Ω, angle of flexion of the forearm; θ, angle of insertion of the brachialis.

The following clockwise torques act on the elbow joint:

$$(B + C)F_1 \sin \Omega \tag{1}$$

$$(B + C + D)F_2 \sin \Omega \tag{2}$$

$$(B + C + D)F_3 \tag{3}$$

The counterclockwise torque is

$$(BT \sin \theta) \tag{4}$$

where T is the unknown tension.

Equating clockwise and counterclockwise torques and regrouping terms, we have

$$T = \frac{[(F_1 + F_2)(B + C) + DF_2] \sin \Omega + (B + C + F)F_3}{B \sin \theta} \tag{5}$$

The force thrusting the ulna against the humerus is expressed as

$$F = \frac{T(B + A \cos \Omega)}{(A^2 + B^2 + 2AB \cos \Omega)^{1/2}} - (F_1 + F_2) \cos \Omega \tag{6}$$

If the weight of forearm and hand is assumed to be approximately 10 lb and to act in a vertical direction on the center of mass of the limb, and when a push button control, such as is used on a crane—also estimated at 10 lb—acts normally on the palm of the hand, then depending on the included angle between forearm and upper arm, the tension in the brachialis muscle will vary from 2150 lb to as low as 170 lb. The thrust exerted on the elbow joint will vary from a high of 2140 lb to a low positive value of 5 lb to a strong negative (joint separation) force of 700 lb. It can be seen (Figure 12) that unless the push button is located so that the included angle between upper arm and the forearm is between 80° and 120°, the combined effects of tension in the muscle and thrust acting on the joint surfaces can create conditions conducive to joint injury.

If the safe range of joint movement has been exceeded, tensions in the muscles will increase rapidly. Likewise, under the same circumstances, dangerous thrust forces develop within the joint. Alternatively, poor workplace layout may also produce "separation" forces of substantial magnitude. These may not injure the surfaces of the synovial linings but can create conditions conducive to severe luxations. Quantitative biomechanical and kinesiological analysis of man–task systems is essential to the protection of the workforce from the deleterious influences of mechanical noxae.

Standard references (18) permit the rapid quantitative assessment of forces and stresses generated in anatomical lever systems by work. Once the analysis has been made, standard reference tables (19) should be consulted to ascertain whether muscles are stressed to excess or are too weak to accomplish the task as planned.

The term "muscular strength" is defined as the maximum tension per unit area that can be developed within a muscle. A maximal effort resulting from the strongest of motivations will yield a tension of approximately 142 psi. However many authors (20–22) agree that under normal working conditions, heavy work would generate approximately

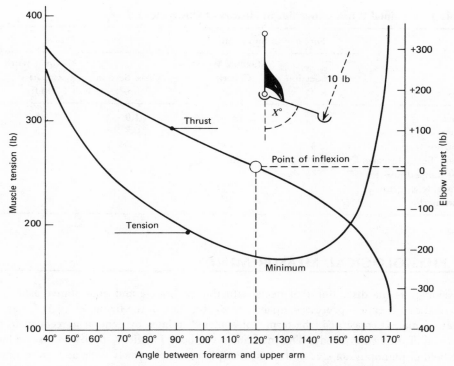

Figure 12 Tension in the brachialis muscle and thrust on the elbow joint generated by a push button requiring 10 pounds of pressure, applied normal to the longitudinal axis of the forearm, for its operation. The combined weight of forearm and hand is assumed to be 10 pounds. (X° = 180° minus Angle.)

50 to 60 psi tension, while light to medium work would produce tension values of the order of 20 to 30 psi. All these values presume maximal shortening of the muscle and are applicable only to continued "ordinary" nonathletic work situations (Table 1).

Occasionally the force available from muscles involved in a kinetic chain is adequate to perform a given task but is unavailable because of a condition known as "muscular insufficiency." A state of active insufficiency exists whenever a muscle passing over two or more joints is shortened to such a degree that no further increase of tension is possible and the full range of joint movement cannot be completed. For example, when a person is seated too low, if the knee joint is hyperflexed it becomes impossible to plantar-flex the foot and operate a pedal. On the other hand, passive insufficiency exists when, in a particular limb configuration, the antagonists passing over one joint are extended to such a degree that it is impossible for the protagonist to contract further. A good example is the inability to close a fist and hold a rod while the wrist is hyperflexed (Figure 13).

Table 1 Maximal Work Capacities of Flexors of Elbow (19–22)

	Forearm in Supination		Forearm in Pronation	
Muscle	Cross Section (in.^{2})	Maximal Work Capacity (lb-ft)	Cross Section (in.^{2})	Maximal Work Capacity (lb-ft)
Brachialis	1.0	28	1.0	28
Biceps	1.1	35	0.8	25
Pronator teres	0.5	9.11	0.65	12
Whole flexor forearm as lumped muscle mass (e.g., grip strength)	3.2	90		

3 PHYSIOLOGICAL MEASUREMENTS

Physiology is the discipline that deals with the qualitative and quantitative aspects of physical and chemical processes intrinsic to the function of the living body. As far as the study of man is concerned, the term "physiology" relates by general agreement to the function of the healthy body. The mechanisms of body functions specific to disease form the field of pathophysiology. Both these disciplines are extremely wide and cover almost all aspects of life or death.

Work physiology is a much narrower field. It is restricted to the effects of work and exercise on physiological function. In industrial practice, work physiologists tend to limit themselves to the study of two narrow aspects of this discipline: the assessments of man's capacity to perform physical work, and the study, as well as description, of the effects of fatigue. Most quantitative results obtained from work physiological evaluation are "output measurements." They constitute physical behavior resulting from the combined effects of a variety of inputs. Oxygen metabolism is not purely a function of severity of exercise or of the lean muscle mass involved; it is also determined, to some degree, by the obesity of the subject, postprandial status, and several emotional factors. The result, however, will be one single number: oxygen uptake expressed in milliliters per minute. Solely on the basis of this result, it will not be possible to make any meaningful statement about the magnitude or quality of the contributing vectors. Therefore most work physiology procedures measure effect but do not permit one to establish cause, except in rare cases of near-perfect technique and experimental design.

The following procedures are commonly employed in industry:

1. Metabolic and quasi-metabolic measurements.
2. Electromyography.
3. The measurement of cardiac performance.
4. The measurement of body temperature and heat loss from the body.

ERGONOMICS

Procedures 3 and 4, however, fall within the purview of the trained specialist and are therefore not discussed here.

3.1 Metabolic and Quasi-Metabolic Measurements

It is common practice in some branches of industry to determine physiological energy expended in the performance of certain tasks by direct measurement or indirect estimation of oxygen consumption. During respiration, oxygen passes through the walls of the pulmonary alveoli and surrounding capillaries into the blood vessels. It is then taken up

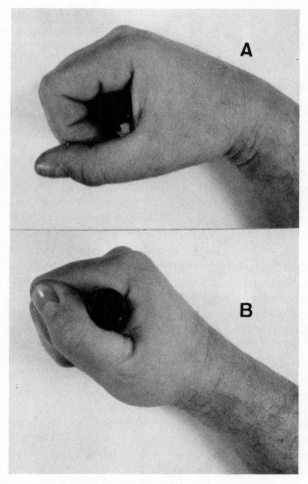

Figure 13 The flexed wrist (a) cannot grasp a rod firmly. (b) The straight wrist can grip and hold firmly. From Reference 48.

by the red blood corpuscles, which are pumped by the heart to body tissues, such as muscles. Oxygen is unloaded in the muscles to take part in physiological combustion processes. The bioenergetics may be represented as follows (23):

At rest

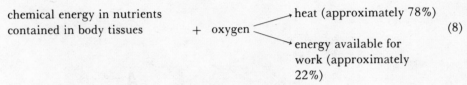

$$\text{chemical energy in nutrients contained in body tissues} + \text{oxygen} \longrightarrow \text{heat} \qquad (7)$$

At work

$$\text{chemical energy in nutrients contained in body tissues} + \text{oxygen} \begin{cases} \longrightarrow \text{heat (approximately 78\%)} \\ \longrightarrow \text{energy available for work (approximately 22\%)} \end{cases} \qquad (8)$$

1. One liter of oxygen consumed releases approximately 5 kcal.
2. 5 kcal ≈ 21,000 J.
3. 4600 J (22 percent of 21,000 J) is available for the performance of work.

Thus energy available per liter of oxygen consumed per minute is approximately 0.1 hp. However ambient fresh air contains approximately 20 percent oxygen, of which, roughly a quarter, or 0.05 liter, becomes available for metabolic activities.

Therefore, as a bench mark figure, it may be assumed that 20 liters of air per minute passing through the lungs over and above normal rest levels of pulmonary ventilation corresponds roughly to 0.1 hp expended in the pursuit of physiological work. This constitutes, however, an extremely crude and not always accurate approximation. The true level of net energy expenditure per liter of air passing through the lungs depends on a wide variety of physiological variables. Yet this has not deterred some industrial enterprises from basing computations of physiological effort, often under conditions of heavy work, on pulmonary ventilation (24). Most commonly, the measurement is performed by a knapsack gasometer, and various models are commercially available (Figure 14). To obtain readings that are of any use at all, it is necessary to measure total airflow through the lungs from the onset of the task to be investigated until the time after termination of work when pulmonary ventilation has returned to resting level. To obtain the net airflow ascribable to the demands of the task, an air volume corresponding to pulmonary ventilation at rest is then subtracted from the total.

More accurate is a procedure developed by Weir (25). In its application, both the knapsack gasometer and an oxygen analyzer are employed. The volume of air passing per minute through the lungs is measured and reduced to the volume corresponding to standard temperature, pressure, and dryness. To compensate for the inaccuracies of the equipment used, this value *must* be further corrected by multiplication with a "calibra-

Figure 14 Metabolic measurement with the knapsack gasometer. Adapted from Reference 80.

tion constant" specific for the actual individual instrument used. The result is obtained as follows:

$$\text{kcal/min} = \frac{1.0548 - 0.0504V}{1 + 0.082d} \times \text{liters vent/min} \tag{9}$$

where V = percentage (vol) oxygen in expired air
d = decimal fraction of total dietary kilocalories from protein

Under most circumstances it is possible to ascertain the protein content of a workman's diet with reasonable accuracy. There is general consensus that this value is relatively constant for each individual.

Methods of metabolic measurement more accurate than those just outlined are complex and sophisticated and should be attempted only by experienced work physiologists. Therefore they are not discussed here. Practitioners active in enterprises where metabolic measurements are taken routinely should familiarize themselves with the basic theory and appropriate techniques through reliable references (26).

3.2 Electromyographic Work Measurement

The aforementioned changes in pulmonary or metabolic activity when measured and quantified by suitable instrumentation can be indicative of the level of effort demanded for the performance of a specific task. Changes in metabolism and pulmonary ventilation represent both dynamic and isometric work performed by muscles; therefore the accuracy of measurement is greatest whenever the musculature of the entire body, or at least large muscle masses such as those of the thighs or the back, must be applied to the successful completion of a job. However in light work, where only mild muscular activity takes place or only small muscle groups or single muscles are utilized, the percentage change in metabolic activity is proportional to the relationship

$$\frac{m}{M} \times 100 = \Delta u$$

where M = total lean muscle mass of body
m = mass of lean muscle applied to task (10)
Δu = percentage change of metabolic activity due to work

In many instances this percentage may be too small to permit a meaningful statement to be made about the level of effort expended, and under such circumstances experimental and computational error may render the result meaningless.

Whenever light work or effort expended by small kinetic chains is to be determined, electromyography can serve as a useful estimator of effort and fatigue.

The term "electromyography," as well as the purpose of the procedure, mean different things to ergonomists, anatomists, physiologists, and physicians. Electrodes, signal conditioning, and display instrumentation vary widely between professions.

In ergonomics, myoelectricity may be assumed to be the "by-product" of muscular contraction which makes it possible to estimate strength and sequencing of muscular activity through techniques noninvasive to the human body. The myogram is an analogue recording of this bioelectric activity.

Each individual muscle fiber maintains in its resting state a negative potential within its membrane wall. This is termed the "resting potential." Excitation produces a transient reversal of the resting potential, causing a characteristic "depolarization" pattern to appear. A discussion of natural and quantitative events related to changes in membrane potential is beyond the scope of this book, and specialized literature should be consulted (27).

Modern electrophysiological thinking assumes (28) that muscle fibers probably never contract as individuals. Instead, small groups contract at the same moment. It has been established that all the fibers in each contracting group are stimulated by the terminal branches of one single nerve fiber, the axon of a motor cell whose body is located in the grey matter of the spinal cord. The nerve cell body proper plus the axon, its branches, and the muscle fibers supplied by them, has been named a "motor unit" (Figure 15). Since an impulse from the nerve cell causes all muscle fibers connected to the axon to contract almost simultaneously, the action potential resulting from such contraction constitutes the elemental event basic to all electromyographic work. This signal can be picked up by a variety of electrodes and amplified and recorded.

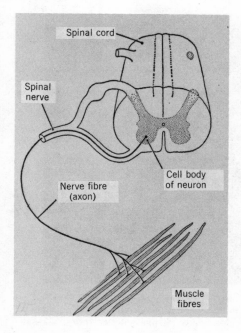

Figure 15 Scheme of motor unit. From Reference 28, by permission. © Williams and Wilkins, Baltimore, Md.

By insertion of very fine needle or wire electrodes into the muscle, it is possible to display action potentials generated by a single motor unit. These individual action potentials, or a sequence of them, are mainly used for electrodiagnostic purposes in medicine. The transducers of choice for electromyographic work measurement are surface electrodes, either permanent or disposable disks of silver coated with silver chloride. More recently, conductive adhesive tape has become commercially available and is often preferred because of economy, flexibility, and ease of application (Figure 16).

Needle electrodes are best avoided in ergonomics. They are invasive, produce pain, and call for great skill in application, moreover, their use entails always a certain risk of infection, particularly under the circumstances prevailing on the shop floor or in the industrial laboratory. Even if the aforementioned difficulties could be overcome, however, the signal generated by needle electrodes is only of limited use to the ergonomist unless he specializes in the more basic aspect of work physiology or electrophysiological kinesiology. The study of single action potentials does not generally permit us to arrive at conclusions relating to the total effort expended by a muscle, and this is why adhesive surface electrodes are preferred in ergonomics. The principal argument against the use of surface electrodes is that unlike needles, they do not permit the study of single action potentials; rather, the signal gathered by them is merely a representation of the level of contractile activity within a relatively large volume of muscle considered to be a "lumped muscle mass" (29). This, however, is the specific

Figure 16 A disposable electrode kit, which is inexpensive, noninvasive, and avoids danger of cross-infection between subjects: *A*, electrode; *B*, adhesive collar; *C*, conductive jelly. The elimination of substantial amounts of time spent on the cleaning of permanent electrodes makes use of this kit very economical.

advantage of surface myography in the study of the activity of whole muscles as opposed to single motor units.

Thus the surface electromyogram may be considered to be representative of the sum of electrical activity generated simultaneously by a large number of motor units. To understand in more detail the theoretical basis of the procedure, the reader should consult specialized reference works (30).

3.3 Electromyographic Technique

In usual practice the myoelectric signal, which is in the microvolt or millivolt range, is gathered by suitably placed electrodes and amplified by a factor of approximately 1000 prior to display by oscilloscope or oscillographic pen recorder.

Myographic apparatus embodying high gain operational amplifiers is generally inexpensive and requires only two electrodes for the production of a myogram. These characteristics make its use more economical, especially where a large number of readings are to be taken. However the operational amplifier, at the levels of magnification involved in myography, produces a noisy signal, often difficult to interpret. Furthermore, a two-electrode system does not permit the investigator to produce repeatable results easily.

ERGONOMICS

Where simplicity of operation and noise-free signals under field conditions are desired, apparatus embodying differential amplifiers is definitely the equipment of choice. A differential amplifier uses three electrodes: one reference, and two active ones. However it augments only the difference between the two active electrodes based on the potential difference between each of them and the reference. Since any interference from external causes will produce identical changes in the potential of all three electrodes, the display signal will not change. This "common mode" rejection makes it imperative to use differential amplifiers in such settings, where electrical interference from fluorescent tubes, radio transmitters, and other equipment is abundant.

In most instances it is desirable (31) to record the myogram by means of one of the numerous commercially available oscillographic recorders, equipped with modular amplifiers and couplers provided by the manufacturer (Figure 17).

Most standard recording equipment has been specifically designed for short-term myography. Whenever data gathering exceeds one hour of operational time, large machines can become uneconomical, and it is recommended that modern miniaturized magnetic tape recorders be employed. These can be easily attached to the belt of a sub-

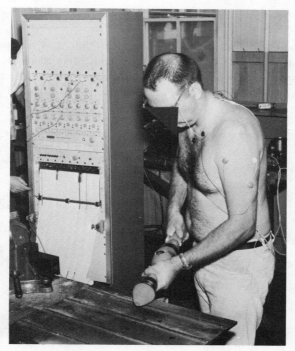

Figure 17 The recording of surface electromyograms by means of a commercially available oscillographic recorder for the purpose of physiological work measurement.

ject, and they record myograms continuously for up to 24 hours on miniature magnetic tape cassettes. In ergonomics the myogram serves three main purposes:

1. To determine the level of effort expended by a specific muscle mass.
2. To determine the nature of sequencing of protagonist and antagonist muscles involved in specific kinesiological maneuvers.
3. To identify or predict localized muscular fatigue.

Correct electroding technique is a prerequisite to successful myography. It starts with muscle testing to determine the location and surface relationships of the individual muscle to be investigated (Figure 18).

By way of example, we discuss the relationship between the biceps and the brachialis muscles. The brachialis is a short and stout muscle that originates from the lower third of the humerus and inserts into the ulna, just distal to the coronoid process (Figure 6). It is a powerful and precise flexor of the forearm (see Section 2.1). More superficially situated, and covering the brachialis, is the biceps. This muscle originates from the scapula and inserts medially into the proximal end of the shaft of the radius (Figure 19).

The biceps is a powerful supinator of the forearm but a comparatively weak flexor. Quite often, especially in materials handling, it is desirable to ascertain the relative magnitude of involvement of either the biceps or the brachialis in the maneuver under study.

These two muscles are located so close together that it is not possible to obtain separate surface myograms for each one unless special precautions are taken. First the brachialis is palpated. This is done by asking the subject to flex the forearm to form an angle of 90° with the upper arm. Then the biceps is pronated strongly against

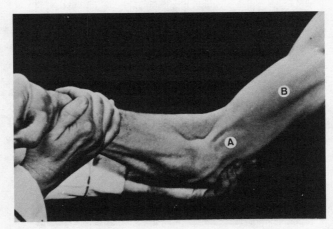

Figure 18 Correct procedure for muscle testing produces the contours and relationships of the biceps *B* as well as the brachialis muscle *A*. Adapted from Reference 32, by permission. © Williams and Wilkins, Baltimore, Md.

ERGONOMICS

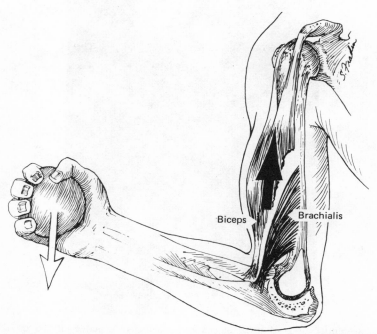

Figure 19 The topographic relationships between two muscles: the biceps and the brachialis. The biceps is superficial to the latter, inserts into the radius, is a powerful supinator but a less effective flexor of the forearm. The brachialis is a short but powerful muscle connecting humerus and ulna; because of the character of the humeroulnar joint as a hinge, this muscle is a very powerful flexor of the forearm. Separate electroding of the two muscles is desirable in electrophysiological work measurement. Adapted from Reference 81.

resistance, while simultaneously the experimenter causes the brachialis to flex the forearm against powerful opposition (Figure 20). The outlines of the muscles are then palpated and the electrodes applied so that a maximum of muscle mass is triangulated by them. The reference electrode is placed conveniently over the triceps tendon. The subject is asked to supinate the flexed forearm strongly against external resistance, and the biceps is palpated and triangulated with an additional set of electrodes (31).

It is essential to provide a separate reference electrode for each muscle investigated. To verify the electrode placement, the subject is first asked to flex the forearm isometrically while the limb is being actively and strongly pronated. This yields a strong brachialis signal but little activity in the biceps (Figure 21). Subsequently the limb is supinated against resistance while being flexed strongly against an opposing force. This produces a strong biceps myogram, concurrent with light to moderate myoelectric activity in the brachialis (Figure 21). Standard reference works (32) should be consulted if it is necessary to acquire proficiency in electroding the muscles contracted during the performance of common industrial elements of work.

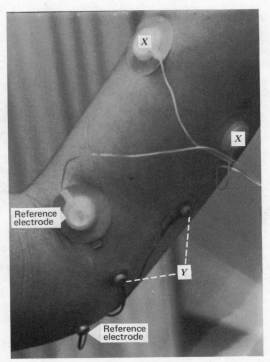

Figure 20 Differential electroding of biceps and brachialis: X, active electrodes of biceps; Y, active electrodes of brachialis.

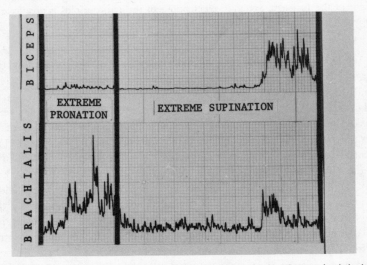

Figure 21 Differential myograms from both biceps and brachialis obtained while hand held 20 pounds of weight; the included angle between forearm and upper arm was approximately 100°.

ERGONOMICS

3.4 Interpretation of Myograms

Once it has been established through muscle testing that the electrodes indeed produce a specific signal, representative of the level of activity in the single muscle under investigation, the real task is performed and the myogram interpreted.

An indispensable prerequisite to the correct interpretation of the signal is an understanding of the circuitry that produces the display and of the operating characteristics of the recording apparatus. Due to the nature of the procedure, myograms produced by surface electrodes are records of the summed signals from a number of action potentials generated simultaneously, near-simultaneously, or consecutively, within the muscle mass triangulated by the electrodes. Since the interval of time elapsing between the peaking of individual action potentials may be as little as a few microseconds, readout devices and pen recorders may be "overdriven." The signal is then not representative of the action potentials but is conditioned and distorted as a function of the quality of the recording device (Figure 22). Even a change in the viscosity of the recording ink may produce a drastic change in the pattern of a tracing from the same amplifier reproduced by the same recorder. Likewise, pen inertia and the quality of maintenance the instrument has

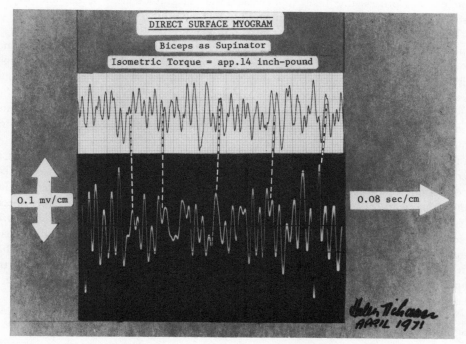

Figure 22 A simultaneous recording of the biceps muscle firing pattern displayed on a chart recorder (*upper half*) and an oscilloscope (*lower half*) at exactly the same sensitivity and speed. Only five points of similarity are evident because the signal speed of the myogram exceeds the rise time and slew rate of commercial chart recorders. From Reference 82, by permission. © American Institute of Industrial Engineers, Inc., Norcross, Ga.

received may cause a badly distorted signal, obfuscating the physiological status of the muscles studied.

To obtain a satisfactory resolution of the direct surface myogram, it is necessary in most instances to run paper recorders at speeds of 12.5 cm/sec. This procedure is uneconomical, and because of friction between pen and paper it gives a completely distorted signal. Therefore in ergonomics a conditioned type of myogram is employed, called the "integrated myogram." However no true integration has taken place. Integrators are circuits that produce a signal that essentially records the sum total of the action potentials counted over a sampling interval of time. This type of myogram is therefore representative of the total number of muscle fibers contracting at any instant. Because of the physiological "all or none" law, it is also representative of the effort expended at any instant. Therefore the area under the integrated myogram is proportional to the total effort made during the time interval under consideration. The generation of this signal requires a much slower recording speed, as low as one centimeter per second, and therefore it can be reliably reproduced by almost any recording device available (Figure 23).

A myogram must not only be read, it must be "interpreted." This entails recourse to a judgmental process that takes into consideration magnitude of pen excursion, general pattern of the tracing, and some features hard to define, such as fuzziness of recording.

The ability to interpret a myogram is an acquired skill, which can be developed within a short time, provided the ergonomist restricts himself, at least in the beginning, to work with tracings obtained from the same make of recorder. This has a twofold advantage. In the first instance, it is easy to develop an appreciation about the "soundness" of the tracing. Second, the integrated surface myogram, the only type of myoelectric readout considered in this context, constitutes a signal conditioned and transformed into shapes and patterns, which may vary for different designs of recording equipment.

Recording speed should be kept constant to facilitate recognition of patterns and to keep slopes and tracings uniform. Many practitioners prefer to work at a recording speed of one centimeter per second. Whenever an exchange of myographic information with other workers is planned, it is advisable for all to adopt the same paper speed. Recorders designed to produce integrated myograms take into consideration both the frequency of peaking of the raw signal and the amplitude of the action potential, but circuits developed by different manufacturers weigh each of these features differently; thus the appearance of myograms from the same muscles, tasks, and individuals, taken at the same occasion, differs considerably according to the make of recorder employed. It is therefore essential not to vary equipment from study to study.

The shape and quality of myograms may also be substantially affected by the following factors:

1. Loose electrodes.
2. Dry electrodes.
3. Loose or broken wiring.
4. Electrical interference from light fixtures or other machinery being used nearby.

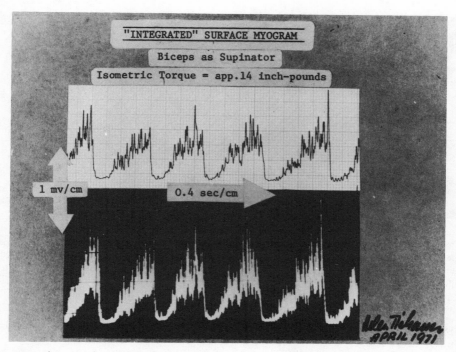

Figure 23 The same biceps contraction pattern shown in Figure 22: chart recorder (*upper half*), oscilloscope (*lower half*). However the signals here have been conditioned by summing all action potentials over a time constant so that the trace now represents the analogue of the firing rate, which is indicative of the total activity of the muscle mass at any instant during the sampling period. The signals are fully compatible with the frequency response of the chart recorder. The "integrated" myogram produces repeatable and very reliable measurements of muscular activity levels. From Reference 82, by permission. © American Institute of Industrial Engineers, Inc., Norcross, Ga.

It is essential to label myograms with the sensitivity settings and speed of the recording device, otherwise tracings obtained at different occasions cannot be compared. Likewise, the baseline should be clearly indicated, or amplitude of pen excursion cannot be quantified. When performing an isometric task (Figure 21), it is relatively easy to ascertain the degree to which each of the muscles investigated participates in the performance of the task, and how changes in musculoskeletal configuration produce a different distribution of work stress acting on the individual members of a kinetic chain. It is also quite simple, when the precondition of isometricity exists, as in static holding (Figure 24), to determine when a critical work stress level has been reached and when a relatively light increase in the severity of the task will produce an undesirably violent myoelectric response.

The area under the integrated myogram has the dimensions of force (volts), multiplied by time (paper speed), which is identical to the dimensions of "linear impulse" in

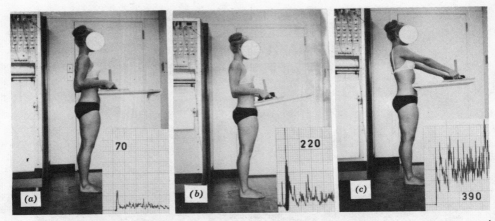

Figure 24 An isometric holding task. Numbers below the myograms represent incremental inch-pounds of torque, applied to the lumbosacral joint, which have elicited the electrophysiological signal. It can be seen that once a certain "critical" level of stress has been exceeded, the electrophysiological signal increases disproportionately as compared with the increment of stress. Under such conditions the subject is at risk. Adapted from Reference 31.

physics and also to the "tension–time" concept used by work physiologists to quantify isometric work (33).

The interpretation of myograms of dynamic tasks, however, is a far more complex matter. The shape of the tracing, its slopes and troughs and amplitudes, are affected by a multitude of factors. These include force and velocity of contraction, tension within the muscle, and whether the contraction is eccentric or concentric.

Thus it is useless in most dynamic situations to even attempt the numeric quantification of the signal. The qualitative discussion, however, can yield information of considerable usefulness. This is the case, for instance, in the analysis of sequencing of different muscles during the performance of a kinesiological maneuver (Figure 25). In wrist rotation the biceps acts as protagonist during supination, while the pronator teres is the antagonist that reverses the movement. Therefore the integrated myograms of both muscles show peak and valley phasing in a nonfatigued, efficiently working individual. When the protagonist fires, the antagonist should be relaxed, and vice versa. In a state of fatigue, however, this clear phasing of muscular activity becomes blurred. A fatigued muscle has lost the ability to relax quickly; therefore the weaker of the two muscles, pronator teres, may fire simultaneously with the biceps, slowing down movement and bringing about undue exertion by the antagonist, increasing further the level of fatigue.

Whenever it is desired to ascertain whether a given musculoskeletal configuration is conducive to an undue expenditure of effort resulting in fatigue or potential trauma, the integrated myogram is the method of investigation of choice. A wire-cutting operation may be deemed to be a quasi-isometric work situation (Figure 26). Very often, when a tool such as a side-cutter is employed while the wrist is in ulnar deviation, the tendons of the flexor muscles bunch against each other inside the carpal tunnel. This produces

ERGONOMICS

friction and necessitates a disproportionately large effort by the muscles of the flexors of the fingers and the thumb for effective performance. This excessive effort becomes immediately apparent on inspection of the myogram.

It is impossible to discuss in detail all the potential uses of electromyographic kinesiology, and specialized papers should be consulted (34–36). However the proper and imaginative use of electromyography constitutes one of the most elegant and useful techniques of ergonomic work measurement.

4 WORK TOLERANCE

Within the context of ergonomics, any action on the living body by any vector intrinsic to the industrial environment is termed "work stress." It is irrelevant whether these vectors are forces and produce movement, whether they merely cause sensory perception or whether, like heat, they increase metabolic activity. All physiological responses to work stress are identified as "work strain." Frequently work stress and the resulting strain occur in anatomical structures quite distant from each other, or the two effects may even

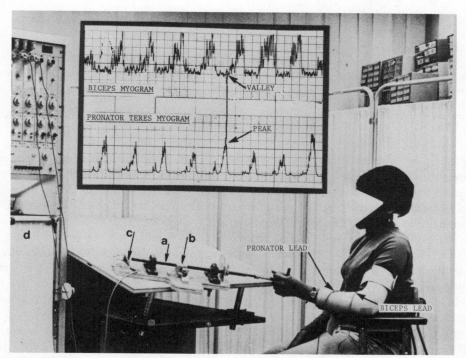

Figure 25 Kinesiometer and subject wired for two surface electromyograms; insert shows how antagonist myograms peak in proper sequence with one another. The kinesiometer consists of A, rotatable shaft; B, friction brake; C, potentiometer; D, recorder. Adapted from Reference 83.

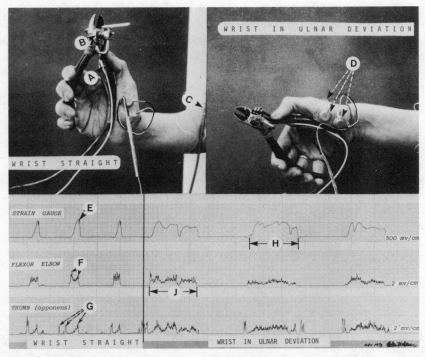

Figure 26 The profile of wire cutting (84). For explanation, see text. A, strain gauge; B, potentiometer; C, electrodes for common flexor myograms; D, electrodes for thenar myograms; E, "notched" strain signal; F, dicrotic flexor myogram; G, multicrotic thumb myogram; H, increased duration of strain signal; J, increased "tension time" of muscle. From Reference 84.

be observed in separate physiological systems. High environmental temperatures are referred to as "heat stress," and the resulting increase in sweating rate is then correctly termed "heat strain." Likewise, in heavy physical work, the forces exerted on the musculoskeletal system are correctly identified as "work stress" and the resulting increase of metabolic activity is an example of "work strain."

The performance of any task, no matter how light, imposes some work stress, and consequently produces work strain in terms of physiological responses. Neither stress nor strain per se is undesirable unless it becomes excessive and diminishes work tolerance.

Work tolerance is defined as a state in which the individual worker performs at economically acceptable rates, while enjoying high levels of emotional and physiological well-being (13). It is common in industrial engineering practice to employ incentive schemes as inducements to increase production rates, thus reducing labor costs. Job enrichment as well as other procedures applied by behavioral and managerial scientists have their place in increasing job satisfaction and in enhancing the social well-being of the workforce. However no management technique available has been found to be successful in overcoming the results of physical discomfort and occupational disease resulting from a poorly designed work situation, ill-matched to the physical operating characteristics of man.

In a room illuminated by a defective spectrum, everyone is color blind. On a job where the motions and reactions inventory demanded by a task is not available from the musculoskeletal system of the worker, everybody is disabled (37), the physically impaired more so. The institution and maintenance of work tolerance has a high priority in the practice of industrial hygiene and ergonomics.

4.1 The Prerequisities of Biomechanical Work Tolerance

The 15 most important prerequisites of biomechanical work tolerance (38) have been arranged in the form of a table (Table 2) and can be employed as a checklist in industrial surveys. Proper use of the table can prevent workplace design from imposing physical demands that cannot be met by a wide range of individual workers. The use of this checklist may also help to avoid the generation of anatomical failure points, which may develop over a number of months, or years, as a result of cumulative work stress. Not all "prerequisites" are applicable to all work situations. However a correctly designed working environment will not violate many of them because this will, beyond doubt, lead to low productivity, poor morale, feelings of ill health and, sometimes, real occupational disease (13).

The "prerequisites" have been arranged in three sets of five statements. The first is concerned with postural integrity; the second relates to the proper engineering of the man–equipment interface; and the third set may be used to ensure that the motions demanded from the workforce are kinesiologically effective.

Table 2 Prerequisites of Biomechanical Work Tolerance

Postural		Engineering		Kinesiological	
P1	Keep elbows down.	E1	Avoid compression ischemia.	K1	Keep forward reaches short.
P2	Minimize moments on spine.	E2	Avoid critical vibrations.	K2	Avoid muscular insufficiency.
P3	Consider sex differences.	E3	Individualize chair design.	K3	Avoid straight-line motions.
P4	Optimize skeletal configuration.	E4	Avoid stress concentration.	K4	Consider working gloves.
P5	Avoid head movement.	E5	Keep wrist straight.	K5	Avoid antagonist fatigue.

4.2 The Postural Correlates of Work Tolerance

P1 Keep the elbows down.

Abduction of the unsupported arm for long intervals may produce fatigue, severe emotional reactions, and also decrements in production rates. The need to keep the unsupported elbow elevated is often the result of poor workplace layout. For example, if the

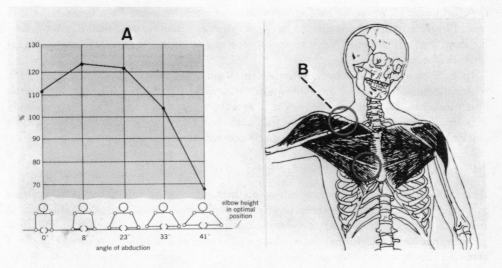

Figure 27 (a) Chair height determines the angle between upper arm and torso, also the moment of inertia of the moving limb. For example, a chair 3 inches too low may raise this moment to a level at which increased effort causes performance to drop to 70 percent of "standard." (b) Also, continued tension in the muscles stabilizing the arm in the raised posture may cause great discomfort at the shoulder over the breast bone. From Reference 37, by permission. © American Institute of Industrial Engineers, Inc., Norcross, Ga.

chair height of seated workers is poorly policed, a seat positioned only 3 inches too low with respect to the work bench will produce an angle of abduction of the upper arm of approximately 45° (Figure 27). When this is the case, wrist movement in the horizontal plane, normally performed by rotation of the humerus, could require a physically demanding shoulder swing. The resulting fatigue over several hours may reduce the efficiency rating by as much as 50 percent. Also, when the seat is too low—especially in assembly operations—the left arm is frequently used as a vise, while the right hand is employed to manipulate objects. This may result in the left arm being held in abduction for several hours. After the elapse of an hour or two, particularly under incentive conditions, some vague sense of discomfort may be felt in the general region of the origin of the left pectoralis major and deltoid muscles, which stabilize the abducted arm. Elderly and overweight workers especially, or individuals with a history of cardiac disease, may develop an unjustified fear of an impending heart attack, and they themsevles, as well as all those around them, may suffer from the ensuing undesirable emotional difficulties (13).

P2   Minimize moments acting upon the vertebral column.

Lifting stress is not solely the result of the weight of any object handled. Its magnitude must be expressed in terms of a "biomechanical lifting equivalent" in the form of a "moment."

ERGONOMICS

The location of the center of mass of the body proper causes a bending moment to be exerted on the axial skeleton even when no object is handled (Figure 28). The muscles erecting the trunk counteract the moment and thus help to maintain upright posture. Thus even simple variations in posture and trunk–limb configuration may modify and substantially increase or decrease, according to the circumstances, the forces exerted on the lumbar spine according to the contribution of the individual body segments to the total moment sum (39). The ergonomic effects of these posture-generated forces are discussed in Section 5. The present checklist is merely concerned with the additional moments imposed on the back by an external load. Very often a light, but bulky object (Figure 29) imposes a heavier lifting stress than a heavy load of great density. It should be remembered that the only way to reduce the lifting stress exerted by an object resides in devising a handling method that will bring the center of mass of the article as close to the lumbar spine as possible.

P3 Consider sex differences.

If employment opportunities for both sexes are to be equal, work environments must be engineered in a manner that takes cognizance of, and compensates for, any sex-dependent differences in anatomy that may affect work tolerance. In the context of lifting tasks, it must be appreciated that male hip sockets are located directly below the

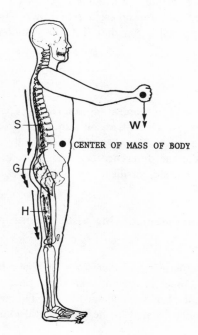

Figure 28 Even when no object is handled, very often a bending moment acts on the vertebral column because of the location of the center of mass of the body. The erector muscles of the trunk counteract this: *S,* sacrospinalis; *G,* glutei; *H,* hamstrings; *W,* weight. From Reference 51.

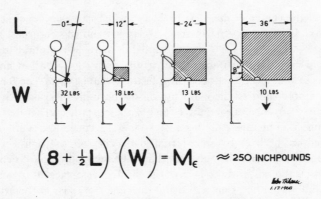

Figure 29 The "moment concept" applied to the derivation of biomechanical lifting equivalents. All loads represented produce approximately equal bending moments on the sacro-lumbar joint, approximately 250 lb-in. Moments exerted by body segments are neglected (13). In the equation, 8 is the approximate distance (inches) from the joints of the lumbar spine to the front of the abdomen, a constant for each individual; L is the length (inches) of one side of the object; W is the weight (pounds) of the object; M_e is the biomechanical lifting equivalent, approximately 250 lb-in. in this example. From Reference 13.

bodies of the lumbar vertebrae; in the female they are situated more forward (Figure 30). A line through the center of the sockets of both hip joints in a woman is located several inches in front of a vertical line passing through the center of gravity of the female body. This activates a force couple. Therefore any object handled by a woman exerts a moment on the back approximately 15 percent larger than if it were handled by a male of identical size or strength.

P4 Optimize skeletal configuration.

Through faulty workplace design, musculoskeletal configuration, especially angular relationships of long bones and muscles, may impose great stress on joints and produce physical impairment (Figure 31). Variations of a few inches horizontally in the distance between the chair and the workspace may make the difference between a productive working population and one that must perform under medical restriction because of great mechanical stress imposed on joint surfaces.

P5 Avoid the need for excessive head movement during visual scanning of the workplace.

It is not possible to estimate correctly, and/or easily, the true sizes or the relative distance of objects except under conditions of binocular vision, which can take place without head movement only within a visual cone of 60° of included angle. The axis of this cone originates from the root of the nose and is located in the mid-sagittal plane of

ERGONOMICS

the head (Figure 32). Head movement at the workplace is often invoked as a "protective reaction" (17) whenever it is necessary to reestablish binocular sight if the visual target is located outside of the cone. Simultaneous eye and head movements take much time, and this may produce a hazard whenever fast-moving equipment—such as motor vehicles, airplanes, or conveyors—are operated. Binocular vision without head movement can be instituted either by dimensioning the workplace appropriately or by changing the position of the operator or the adjustment of the working chair.

4.3 The Engineering of the Man–Equipment Interface

E1 Avoid compression ischemia.

The term "ischemia" describes a situation in which blood flow to the tissues is obstructed. It is essential that the designer as well as the evaluator of tools and equipment be familiar with the location of blood vessels vulnerable to compression. Improperly designed or misused, a piece of equipment may have the effect of a tourniquet. Of special importance is a knowledge of the location of blood vessels and other pressure-sensitive anatomical structures in the hand. A poorly designed or improperly

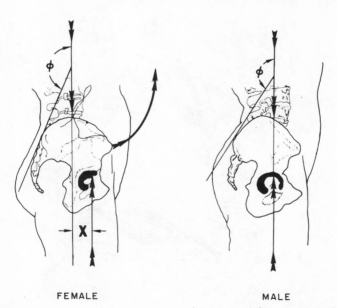

Figure 30 Hip sockets in the male are located directly under the bodies of the lumbar vertebrae in the same plane as the center of mass of the body. In the female, the sockets are located further forward, represented by the distance X. This produces a force couple so that the lifting stress in the back muscles in women, for the same object, can be as much as 15 percent higher than in males. From Reference 85, by permission. American Industrial Hygiene Association, Akron, Ohio.

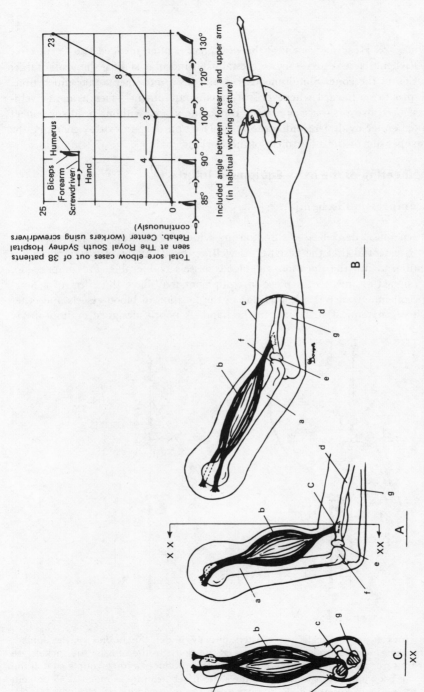

Figure 31 The mechanical advantage of the biceps depends on the angle of flexion of the forearm. This muscle is not only a flexor, but also, because of the mode of attachment, the most powerful outward rotator of the limb. The worker who sits too far away from his workplace has to overexert himself when using a screwdriver because the biceps operates at mechanical disadvantage. Sore muscles and excessive friction between the bony structures of the elbow joint are the results (80). (A) The forearm flexed at 90 degrees: *a*, humerus; *b*, biceps; *c*, attachment of biceps; *d*, radius; *e*, head of radius; *f*, capitulum of humerus; *g*, ulna. (B) The angle of the forearm extended when the efficiency of the biceps as an outward rotator is reduced. Here the muscles will pull the radius strongly against the humerus, causing friction and heat in the joint. (C) Cross section X-X through A, showing why the biceps is an outward rotator of the forearm. From Reference 80.

ERGONOMICS

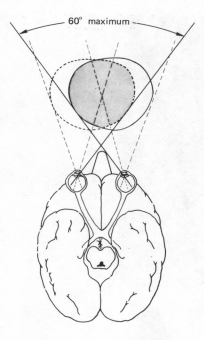

Figure 32 Eye travel and binocular vision. Whenever an object is located outside a binocular field of vision (shaded area) depth perception is impossible; head movement will automatically ensue to correct the deficiency. Heavy dotted lines indicate convergence at 16 inches. From Reference 48.

held hand tool may squeeze the ulnar artery (Figure 33). This may lead to numbness and tingling of the fingers. The afflicted worker will put down his tools and may make use of any reasonable excuse to absent himself temporarily from the workplace as the only means of relief open to him. Apart from the resulting drop in productivity, the health of the working population under such circumstances is in serious jeopardy inasmuch as cases of thrombosis of the ulnar artery and other instances of permanent damage have been reported. Unless the engineering and medical departments are alerted to a possible mismatch between hand and tool, the complaint of numb and tingling fingers could be erroneously attributed to one of the numerous other causes of these symptoms, and the sufferer improperly diagnosed or treated (41). Generally speaking, handles of implements should be designed to make it impossible for the tools to dig into the palm of the hand or to exert pressure on danger zones.

E2 Avoid vibrations in critical frequency bands.

Vibrations transmitted at the man–equipment interface can easily lead to somatic resonance reactions. "White finger syndrome" or intermittent blanching and numbness of the fingers, sometimes accompanied by lesions of the skin, has been identified for many years as an occupational disease associated with the operation of pneumatic hammers and other vibrating tools (42, 43). It is cited as only one example of numerous diseases and injuries caused by exposure to vibrations. When exposed to critical ranges of vibra-

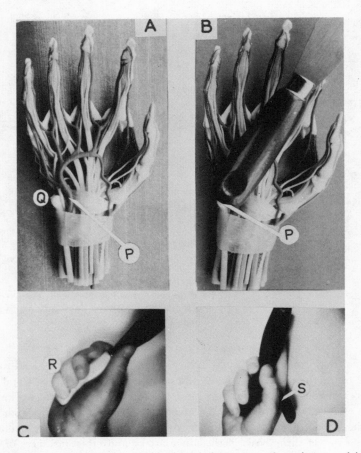

Figure 33 Ergonomic considerations in hand tool design. (a) The relations of bones, blood vessels, and nerves in the dissected hand. (b) A paint scraper is often held so that it presses on a major blood vessel P and directs a pressure vector against the hook of the hamate bone Q. (c) In the live hand, this results in a reduction of blood flow to, among others, the ring and little fingers, which shows as light areas on infrared film R. (d) A modification of the handle of the paint scraper causes it to rest on the robust tissues between thumb and index finger S, thus preventing pressures on the critical areas of the hand. From Reference 40, by permission. American Industrial Hygiene Association, Akron, Ohio.

tion, various viscera, muscle masses, and bones may react in an undesirable manner. This can simulate a wide range of diseases that are commonly associated with musculoskeletal discomfort, including back pain, respiratory difficulties, and visual disturbances. The critical ranges of vibration that may produce such undesirable side effects are fortunately very narrow, and it is often easy, through recourse to such normal engineering procedures as construction of vibration-absorbing tool handles, to avoid exposure to noxious frequencies. An "epidemic" of otherwise inexplicable afflictions of the musculoskeletal system should always alert the ergonomist to consult a reliable reference work on industrial vibration (44).

E3 Individualize chair design.

The design of any seating device should match the need of individual work situations. The anthropometric and biomechanical basis of chair design and adjustment are discussed in section 7. When evaluating work situations on the shop floor, it should be remembered that universally useful "ergonomic work chairs" do not exist. Likewise, diagrams that show "standard dimensions" of the components of a work chair, or of seating height, are highly suspect unless they list a wide range of tolerances for each dimension. By way of example, it is mentioned that the pilot seat in an aircraft should support the trunk while the pilot is sitting still. On the other hand, the chair on an assembly line should give adequate lumbar support but at the same time permit the body to perform all necessary productive movements (45) (Figure 55). Working chairs should have an adjustable-height backrest that swivels about a horizontal axis, to be able to adapt to the demands of the contours of the back. It should be small enough not to interfere with the free movement of the elbows during work. The seat should be adjustable in height, and it should be complemented by an adjustable footrest to relieve pressure exerted by the edge of the seat on the back of the thigh.

E4 Avoid stress concentration on vulnerable bones and joints.

Sometimes features in tool and equipment design look deceptively advantageous but are in reality most dangerous to the integrity of the skeletal system. Finger-grooved tools are an example (Figure 34). They fit perfectly one hand—the hand of the designer. If gripped firmly by a hand too large or too small, the metal ridges of the handle may exert undue pressure on the delicate structures of the interphalangeal joints (46). This will make it painful to grip the tool firmly. Sometimes worse, the working population may show signs of discomfort and absenteeism and will be exposed to medical restriction of performance levels. Finally, under the worst of circumstances, permanent and disabling bone and joint disease may result. Sometimes simple shielding devices are quite effective to protect anatomical structures from stress concentration. The tailor's thimble is a good example of this—it is, perhaps, the oldest device in history protecting tissues against stress concentration.

E5 Keep the wrist straight while rotating forearm and hand.

Four wrist configurations, particularly when they approach the extremes of their range, are conducive to fatigue, discomfort, and sometimes disease. These are: (*a*) ulnar deviation, (*b*) radial deviation, (*c*) dorsiflexion, and (*d*) palmar flexion.
  Especially unhygienic situations are those in which these positions alternate fairly rapidly during the work cycle, or occur in combination with each other. Unfortunately

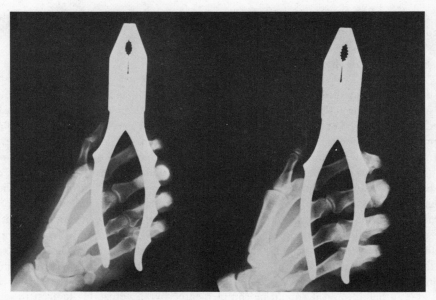

Figure 34 Form-fitting grips on hand tools may cause severe pressures on the finger joints of a person with a larger hand than the hand size for which the tool was designed. From Reference 46, by permission. © Journal of Occupational Medicine, Chicago, Ill.

tools and workplaces are often so designed that the aforementioned movements are demanded as part of the normal work cycle. This affects both health and efficiency. The principal flexor and extensor muscles of the fingers originate in the elbow region, or from the forearm, and are connected to the phalanges by way of long tendons. The extensor tendons are held in place by the confining transverse ligament on the dorsum of the wrist, and the flexor tendons on the palmar side of the hand pass through the narrow carpal tunnel, which contains also the median nerve.

Failure to maintain the wrist straight causes these tendons to bend, to become subject to mechanical stress, and to traumatize such ancillary structures as the tendon sheaths and some ligaments. Ulnar deviation, combined with supination, favors the development of tenosynovitis. Radial deflexion, particularly if combined with pronation, increases pressure between the head of the radius and the capitulum of the humerus. This is conducive to epicondylitis or epicondylar bursitis. These conditions are frequently observed in those who operate hand tools. If hand tools like screwdrivers or pliers are pronated and supinated against resistance for only a few minutes during the day, no harm results. However continuous production jobs may constitute a hazard to the working population. Generally speaking, it is safer to bend the implement than to bend the wrist. Furthermore, both ulnar and radial deviations of the wrist, when combined with simultaneous pronation or supination, reduce the range of rotation of the forearm by more than 50 percent (46). Under these circumstances, the afflicted individuals are obliged to go through double the number of motions to perform a given task, such as

ERGONOMICS

looping a wire around a peg (i.e., they have to work twice as hard for half the output). This may dramatically increase personnel turnover and lead to massive dropouts of new workers during training (13) (Figure 35).

4.4 The Development of Effective Kinesiology

K1 Keep forward reaches short.

Numerous industrial engineering texts describe and define "normal" and "extended" reach areas, specifying that the motion elements "reach" and "transport" may safely be

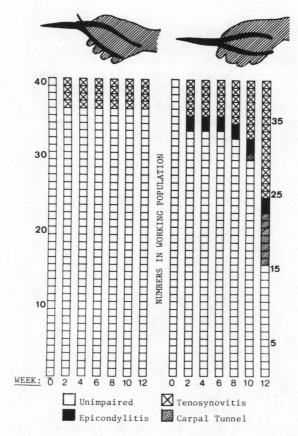

Figure 35 Comparison of two groups of trainees in electronics assembly shows that it is better to bend pliers than to bend the wrist (37). With bent pliers and wrists straight, workers become stabilized during second week of training. With wrist in ulnar deviation, a gradual increase in disease is observed. Note the sharp increase of losses in tenth and twelfth weeks of training: only 15 of 40 workers in the sample remained unimpaired. From Reference 37, by permission. ©American Institute of Industrial Engineers, Inc., Norcross, Ga.

included in repetitive tasks for continuous work provided they do not exceed 25 inches (47). This assumption is fallacious. The protagonist muscles of forward flexion of the upper arm operate at biomechanical disadvantage. One of their principal antagonists is the large and powerful latissimus dorsi (Figure 36). Whenever an extended and fast reach movement in the sagittal plane away from the body is performed, a strong stretch reflex is produced in the latissimus dorsi. This, in turn, subjects the vertebral column to compressive and banding stresses. Frequent and rapid forward reach movements, especially when associated with the disposal of objects, require that the length of this motion be kept to somewhat less than 16 inches. The operative word here is "frequent." Extended reaches per se, especially when performed slowly and relatively rarely during the working day, are quite harmless.

K2 Avoid muscular insufficiency.

Sometimes a specific muscle is unable to produce an expected full range of joint movement. Such a situation is defined by the term "muscular insufficiency" (15). It occurs when the protagonist is contracted to such a degree that it cannot further shorten, or an antagonist has become hyperextended and impedes further joint movement. By way of example, when the wrist is fully flexed, the hand cannot grasp a rod firmly because the extensors of the fingers are overextended, and the flexors are overcontracted (Figure 37).

Figure 36 The disposal of objects in a direction away from the body and in the mid-sagittal plane is conducive to discomfort and early fatigue because of the strong antagonist activity in the latissimus dorsi muscle. From Reference 48.

ERGONOMICS

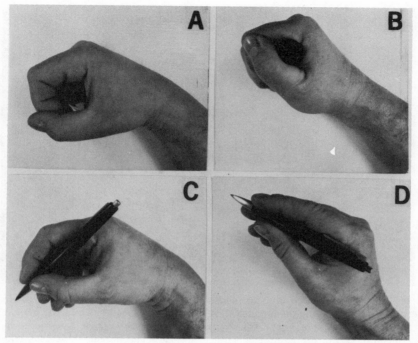

Figure 37 Flexed wrist (A) cannot grasp a rod firmly; the straight wrist (B) can grip and hold it firmly. Conversely, the flexed wrist (C) is well positioned for fine manipulation, but when extended (D), freedom of finger movement is severly limited. From Reference 48.

K3 Movements along a straight line should be avoided.

All joints involved in productive movements at the workplace are hinges. Therefore the pathway of an anatomical reference point is always curved when it results from the simple movement of a single joint. Such curved movements can be produced by a contraction of one single muscle. On the other hand, motion along a straight line requires higher skills not always available from a subclinical impaired working population (e.g., the aged). Straight line movements generally require longer learning, produce early fatigue, and are less precise than simple ballistic motions (48).

K4 Working gloves should be correctly designed.

Often occupational hazards and operational inefficiencies are erroneously ascribed to improper tool or usage when, instead, the working glove is implicated. The end organs of the sensory nerves of the hand are, among others, also distributed along the interdigital surfaces of the fingers (49) (Figure 38). Since it is impossible to close the fist

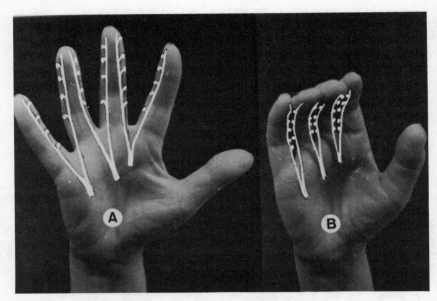

Figure 38 Nerve endings that provide feedback information about the degree of closure of the hand are located between the fingers (A). When the hand executes a gripping motion (B), the fingers abut and the nerve endings press against each other. Work gloves that are too thick may produce pressure against the interdigital surfaces too early and provide misleading information about the firmness of grip when holding a heavy or slippery object. From Reference 37, by permission. © American Institue of Industrial Engineers, Inc., Norcross, Ga.

firmly while the fingers exert pressure against each other, it is difficult to maintain grip strength under such circumstances. If a working glove is too thick between the fingers, high pressure on the interdigital surfaces may be generated before the hand is firmly closed about a tool handle or equipment control. This results in an insecure grasp. Lack of awareness of this lowered capability may cause heavy objects to slip out of the hands of workers, resulting in accidents. It may also lead to inadequate control over cutting tools or dials operated in a cold climate.

K5 Antagonist fatigue should be considered in task design.

In the performance of many simple movements, the muscles that reestablish the initial condition after a specific maneuver has been performed are often weaker than those which bring about the primary movement. For example, when inserting screws by means of a ratchet screwdriver, the biceps is the outward rotator of the forearm, and this strong muscle will not easily fatigue even when operating against strong resistance. Even when operating against a resistance too small to be measurable, however, the opposing inward rotator of the forearm, the pronator teres, fatigues easily because of its small size (19). Physiological work stress should never exceed the capacity of the smallest muscle involved in a kinetic chain. Reliable tables relating name of muscle, function, cross section, and working capacity should be consulted by the practicing ergonomist (15).

The 15 prerequisites of biomechanical work tolerance just given are no substitute for a comprehensive knowledge of the theory and practice of ergonomics. They are merely a convenient expedient for the rapid and gross evaluation of work situations during discussions on the shop floor. Properly employed in discussions with first-line supervisors and engineering personnel engaged in the planning and design of work situations, these prerequisites can help to reduce, without access to complex esoteric and costly laboratory facilities, the incidence of occupational accident and disease right where it occurs: on the shop floor.

5 MANUAL MATERIALS HANDLING AND LIFTING

Almost one-third of all disabling injuries at work—temporary or permanent—are related to manual handling of objects (48). Many of these incidents are avoidable and are the consequence of inadequate or simplistic biomechanical task analysis. The relative severity of materials handling operations, and differences in lifting methods, can be evaluated only when all elements of a lifting task are considered together as an integral set (Table 3). All these elements have different dimensional characteristics, but nevertheless have one basic property in common: any major change in magnitude of any element of a lifting task produces a change in the level of metabolic activity.

No matter what the dimensions of mechanical stress imposed on the human body during materials handling, the physiological response will always result in changes of energy demand and release, customarily expressed in kilocalories. Thus physiological response to lifting and materials handling stress has always the dimensions of work. Hence when the task is heavy enough, the measurement of metabolic activity provides a

Table 3 The Elements of a Lifting Task (48)

Static Moments	Gravitational Components	Inertial Forces
Sagittal	Isometric	Acceleration
Lateral	Dynamic	Aggregation
Torsional	Negative	Segregation

Frequency of task

convenient experimental method for the objective comparison of the relative severity of materials handling chores.

Current consensus (26) assumes on the basis of an 8-hour working day that the limit for heavy continuous work has been reached when the oxygen uptake over and above resting levels approaches 8 kcal/min. The upper limit for medium-heavy continuous work, seems to be 6 kcal/min, and an increment of 2 kcal/min appears to be the dividing line between light and medium-heavy work. However in many instances the application of metabolic measurement is not feasible. The complex procedure may be too difficult to perform on the shop floor. Furthermore, the assessment of work stress through the analysis of respiratory gas exchange is, by definition, an ex post facto procedure. The job exists already, and the energy demands are merely computed to decide whether corrective action is indicated. It is, of course, much better to analyze a task objectively while both job and workplace layout are still in the design stage. Recourse must then be taken to "elemental analysis."

5.1 Elemental Analysis of Lifting Tasks

Any activity producing a moment that acts—no matter in which direction—on the vertebral column, must be classified as a "lifting task." A "moment" is defined as the magnitude of a force multiplied by the distance from the points of its application. In most instances ergonomic analysis of lifting tasks is concerned with moments acting on the lumbar spine or, when a specific task involves head fixation, with moments acting on the cervical spine.

There are three static moments to be considered (Table 3). In many instances they are easily computed by direct measurement or with the aid of drawings or photographs. Sometimes, however, it is more convenient to estimate them by speculative analysis or visual inspection of the work situation. They are conveniently expressed in pound-inches (lb-in.), or kilogram-centimeters (kg-cm). This value is obtained by multiplying the force acting on an anatomical structure with the distance from the point of maximal stress concentration.

The heaviest article normally handled by man at work is his own body, or its subsegments. Only rarely do workers handle objects weighing 150 pounds and, in most instances, the mass of an object moved is quite insignificant when compared with the weight of the body segment involved in the operation. For example, the majority of hand tools or mechanical components in industry weigh considerably less than 0.5 pound, but an arm, taken as an isolated body segment, weighs 11 pounds (50).

The *sagittal lifting moment* is the one most frequently encountered and easiest to compute. It is most conveniently derived by graphical methods. First the weights of the body segments involved in a specific task are obtained from reliable tables (18). Then a "stick figure" of proper anthropometric dimensions (Figure 39) is drawn, and the location of the center of mass for each body segment is marked, as well as the center of mass for the load handled. Finally, the sum of all moments acting on the selected anatomical reference structure (in this case, the lumbosacral joint) is computed and becomes the

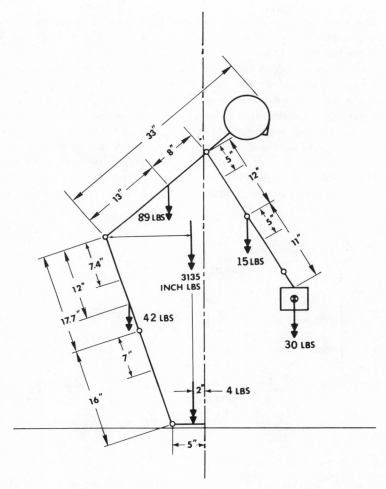

Figure 39 Graphic computation of the location of center of mass of whole body and body segments as well as of the sagittal moment acting on the lumbosacral joint can be conveniently accomplished through the use of stick figures. This example shows that in improper working posture, a load weighing only 30 pounds, combined with the mass of the various body segments involved in a lifting task, may produce a torque exceeding 300 in.-lb, which is the lifting equivalent of a very severe task. Illustration courtesy of Dr. C. H. Saran; from Reference 48.

sagittal biomechanical lifting equivalent of the specific task under consideration. Whenever the vector representing the moment sum of all gravitational moments is directed at a point on the floor located in front of or behind the soles of the feet, a prima facie hazard exists because of inherent postural instability and the ensuing likelihood of falls.

The estimation of sagittal lifting equivalents is of great practical usefulness in the comparison of work methods or in the assessment of the relative magnitude of lifting stress due to sex differences (Figure 40).

Males and females of approximately the same height and weight may be subject to different stresses when handling the same object. This is because of sex-dependent differences of the relative proportions of body segments. Moments acting on the lumbosacral joint during lifting depend largely on work surface height; therefore females are often at a disadvantage when picking up objects from the floor, but a pallet 12 to 14 inches high reduces sex differences in lifting stress during load acquisition (51).

Sometimes it is necessary to decide whether a task would be better performed sitting, as opposed to standing (Figure 41). Then a sketch, true to scale, or a photograph, becomes a convenient aid to decision making. If the pictorial representation of the work

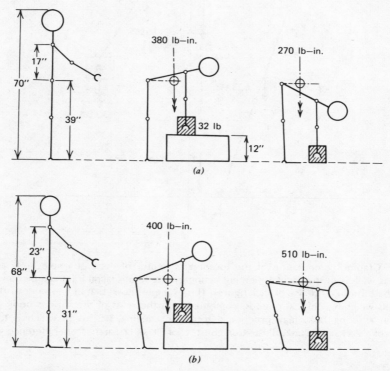

Figure 40 (a) Males and (b) females of approximately the same height and weight may be subject to quite different stresses when handling the same object. This is due to sex-dependent differences of the relative proportions of body segments. Because moments acting on the lumbosacral joint when lifting depend largely on work surface height, females are at a disadvantage in certain postures during load acquisition, whereas in others sex differences are minimal. Adapted from Reference 51.

ERGONOMICS

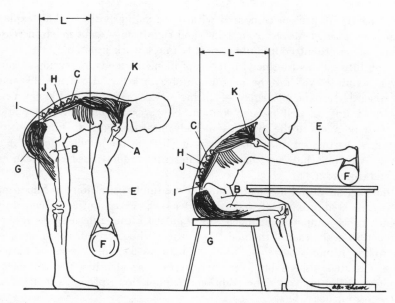

Figure 41 Awareness of the "hidden" lifting task should exist. Because of increase of dimension L, hence higher torques exerted on the lumbar spine, a seated job, instead of being "light work" may be the physiological equivalent of a severe lifting task. In seated work, the rule is: "get the job close to the worker." A, humerus; B, socket of hip joint; C, vertebral column; D, shoulder joint; E, arm; F, load; G, muscles of the buttocks (gluteus maximus); H, muscles of the back (sacrospinalis); I, lumbosacral joint; J, spinous process of a vertebra; K, trapezius muscle; L, distance from the center of mass of combined body–load aggregate to the joints of the lumbar spine. From Reference 58, by permission. © American Institute of Industrial Engineers, Inc., Norcross, Ga.

situation is supplemented by the estimated weight of the body segments involved and the location of the respective centers of gravity, the relative lifting stress for both postures can be easily estimated. Since the procedure does not compare lifting stresses between different individuals but merely establishes how the same person is affected by changes in posture, it may be assumed that the body segments involved (i.e., the torso above the lumbosacral joint, plus neck, head, upper limb, and the object manipulated) are identical in both the seated and standing postures.

According to standard data (52) the body mass in the case of a 110-pound female would be 45 pounds. To this must be added the weight of the object handled—20 pounds in the example under consideration. For this example, the distance from the lumbosacral joint to the center of mass of body segments and load combined may be estimated to be 1½ feet when standing. This exerts a torque of approximately 98 foot-pounds (ft-lb). However if the individual is seated, this value increases to approximately 2½ feet because of the forward leaning posture of the trunk and the outstretched arm. Therefore the torque now exerted on the lumbar spine amounts to 146 ft-lb, or an increase of nearly 50 percent compared with the standing position. This explains why, in so many instances, emloyees complain—and rightly so—about much increased work stress when chairs are introduced unnecessarily into a work situation.

Analyzing lifting tasks routinely in terms of moments tends to develop in supervisors not only a "clinical eye" for the magnitude of a task but also a healthy and critical attitude toward "cookbook" rules of lifting. The principle of "knees bent—back straight—head up" is well enough known. However in many work situations concessions must be made to the influence of body measurements. In Figure 42 the male, long legged, short torsoed, does not benefit at all from the application of the standard lifting rule. A female, however, having a differently proportioned body, can get under the load and close to it.

Thus the sagittal lifting moment acting on the lumbosacral joint becomes much less, and work stress is approximately halved. Under such circumstances, provided the height of the work bench cannot be changed, the standard lifting rule may be applied to the female, whereas in the case of the male no benefit will be derived. Working according to the "approved" lifting posture could lull the male worker into a sense of false security. Therefore it is always advisable to temper the categorical instruction of "knees bent—back straight—head up" with the additional explanation "provided it helps to get the load closer to your body."

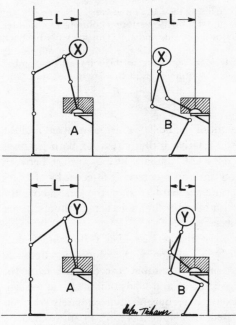

A "WRONG" POSTURE
B "APPROVED" POSTURE

Figure 42 Postural corrections in training for lifting should be aimed at reducing torques acting on the spine: X, an anthropometric male, does not benefit materially from the "approved" lifting posture because L, the distance from the center of mass of load to the fourth lumbar vertebra, does not shorten materially; Y, an anthropometric female, does benefit from the "bent knees, straight back" rule because she can get under the load. When matching worker and task, the measurements of the individual worker as well as the dimensions of the workplace should be considered. From Reference 58, by permission. © American Institute of Industrial Engineers, Inc., Norcross, Ga.

ERGONOMICS

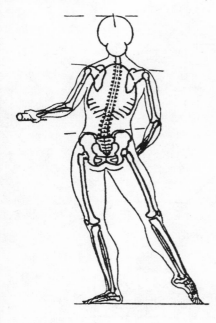

Figure 43 Sidestepping induces heavy lateral bending moments acting on the spine (86).

It has already been described (Figure 29) how a light but bulky object often imposes a lifting stress much greater than the one exerted by a heavier article of greater density. A metal ingot held close to the body exerts a lesser sagittal bending moment on the spine than a box of equal weight containing small miniaturized components, such as transistors packed for shipping in a Styrofoam container. This age of miniaturization and containerization has added a serious and sinister overtone to the age-old joke: "Which is heavier, a pound of lead or a pound of feathers?" The feathers, of course; they are so much bulkier.

In all instances, however, the two other moments in addition to the sagittal one, must also be considered. *Lateral bending moments* are all-important whenever a job calls for "sidestepping" (Figure 43). This often occurs when, for example, the serving of food at lunch counters is involved, or when components have to be transferred to a tray from a jig mounted on a machine.

Generally speaking, bench work involves unnecessary sidestepping. Lateral bending moments can be of considerable magnitude, and in special cases (e.g., when an individual suffers from a mild nerve root entrapment syndrome), they may impose considerable hazard and suffering. A workplace designed for sidestepping is a workplace designed for trouble.

Consideration of *torsional moments* acting on the vertebral column becomes necessary when materials are transferred from one service or work bench to another (Figure 44). The "L"-shaped work surfaces appeal to both architect and industrial engineer because they combine aesthetic considerations with opportunities for performance efficiency.

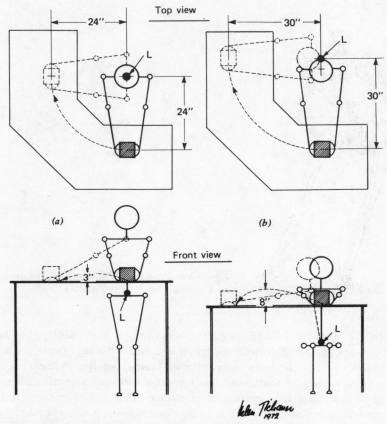

Figure 44 The schematic drawing shows that because of relative fixation of the pelvis, a seated lifting task may produce very heavy torsional bending moments acting on the lumbar spine; L is the lumbosacral joint. From Reference 48.

Whenever a lifting task requiring rotation of the torso about the vertical axis of the body is to be performed by a standing person, a serious hazard is not presented because of the interaction between ankle, knee, and hip joints. However when such a task is performed seated, the pelvis is securely anchored and the entire torsional moment must be absorbed by the lumbar and thoracic spine. This condition is conducive to great mechanical stress on the vertebral column; it may aggravate preexisting back pain caused by pathology in the lumbar or lumbosacral region, and it can cause distress, sometimes of a respiratory nature, to scoliotic or kyphotic individuals.

It is easy to avoid excessive torsional moments in seated work situations by providing well-designed swivel chairs. It is possible to determine numerically the severity of moment action on the human body during a lifting task by the use of computerized models (53). These have been developed to a fairly high degree of perfection, and their use constitutes the procedure of choice in mass production operations under conditions of rigidly controlled work methods. One of the principal advantages of the computerized lifting models resides in their capability to permit accurate estimation of moments exerted by the mass of various body segments on a wide variety of anatomical reference points. Sometimes, however, this accuracy and convenience must be sacrificed when the need occurs to make rapid ad hoc decisions under field conditions or when the small size of the workforce makes recourse to a large computational facility uneconomical.

It is for this reason that ergonomists should aim to develop the knack of "guesstimating" the magnitude of all three moments by looking at the worker, at motion pictures of an operation, or even at a drawing of the workplace layout. Then these guessed moments should be added, not algebraically, but vectorially.

By way of bench mark, it may be assumed that when the vector sum of all three moments is 350 in.-lb or less, the work is light and can be performed with ease by untrained individuals, male as well as female, irrespective of body build. Moments above this level, but below 750 in.-lb, classify a task as "medium-heavy," requiring good body structure as well as some training. Tasks above this moment but below 1200 in.-lb may be considered to be heavy, requiring selective recruitment of labor, careful training, and attention to rest pauses. Whenever the vector sum of moments exceeds those stated before, the work is very heavy, cannot always be performed on a continuing basis for the entire working day, and requires great care in recruitment and training.

It is my personal experience that the ability to guess the magnitude of moments exerted by the load on the body is easily acquired. The same does not hold true, however, for moments exerted by segments of the living body. Therefore the validity of the aforementioned bench mark values is somewhat limited because they do not take into consideration individual differences in lifting stress caused by body segments, but refer to moment increases caused by external loads only.

The *gravitational components* are elements of a lifting task closely related to the concept of "work" in the sense of physics. In mechanics, work is defined as the product of force multiplied by the distance through which it acts. Thus lifting 10 pounds against gravity to a height of 5 feet will constitute 50 ft-lb of work. Likewise, pushing an object horizontally for a distance of 5 feet when 10 pounds of pushing force is required will also result in 50 ft-lb of work. This definition, however, is not always applicable to a situation involving the human body at work. For example, if an individual pushes with all his force against a wall and moves neither his body nor the wall, he has not accomplished any work according to the rigid definitions of physics. Nevertheless, during the entire time, muscles have been under tension, metabolic activities increased, and the added energy demands of the living organism manifested themselves in the expenditure of additional calories which, in physics, are assigned the dimensions of work.

The effort expended on an activity that requires the application of force for a period of time without concurrent displacement of an object is called "isometric activity." Sometimes the term "tension time" is also applied to this kind of activity. "Isometric work" is assigned the physical dimensions of "impulse," which equals force multiplied by time. This makes mathematical processing somewhat difficult inasmuch as gravitational components of a lifting task have the dimensions of "work," which equals force multiplied by distance. Mathematical transformations to overcome this difficulty have been developed (54) and are useful whenever recourse to computerized models must be taken. For all practical purposes, it is often adequate to estimate *isometric work* by taking the weight of the object handled plus the estimated weight of the body segment involved in the task, and to multiply these by the time the muscles are under tension.

Dynamic work is defined as the product of the weight of an object handled, multiplied by the vertical distance through which it is lifted upward against gravity. It has the dimensions of work as defined in physics and can be generally computed with ease.

Negative work is performed whenever an object is lowered at velocities and accelerations of less than gravity so that work against the "g" vector is performed.

To avoid complex computations, it is practical to assume, under industrial working conditions (55), that one-third of the work that would have been expended when lifting the same object over the same distance in an upward direction is approximately equal to the negative work performed.

Finally, in the evaluation of a lifting task, the *inertial forces* must be considered. When an object handled is in motion, acceleration is generally insignificant as far as work stress is concerned. However the forces involved in *aggregation* and *segregation* of man and load may impose severe stresses on the human body.

To maintain the unstable equilibrium of upright posture, it is necessary that the center of mass of the body be located over a line connecting the sesamoid bones of the big toes. Whenever a load is lifted, object and human body become one single aggregate, and as soon as the load has left the ground, the body, through changes in postural configuration, must place the center of mass of this body–load aggregate over the area of support. This requires displacement of the center of mass of the body proper which, during normal acquisition of the load from the ground, takes place over a time interval of roughly 0.75 second. During this brief time interval "stress spikes" will be observable if myograms of the muscles of the back are taken (56). Acquisition stress can best be minimized by having the point of pickup of the load as close to the worker's body as possible.

A stress far more severe is experienced on segregation of the load. Since release of an object is normally fairly rapid, segregation may take place over as small an interval of time as 40 msec, or $1/20$ of the time involved in acquisition. This requires that postural adjustments be far more rapid during release than during pickup, which, in turn, produces great stress on the musculoskeletal structures involved. Electromyographic studies (31) have shown that stress experienced during release may be a multiple of the physical work stress generated during the rest of the work cycle. This has been confirmed by other investigators (56). Segregation stress can best be reduced by having the point of release of the load as high above floor level as possible. It is also essential that workers be made aware of the postural adjustments occurring during load release, including pelvic rotation, and be trained to perform these slowly.

Finally, *frequency of lift* is often one of the elements deserving considerable attention. The number of times a lifting task is performed during the day is equivalent to "productivity," and therefore it is determined by economic needs. Where bulky loads, such as television tubes, are handled, it may be quite impossible to control the severity of a lifting operation by attention to the frequency of lift, since this will not be amenable to modification. Yet often a number of relatively small units, several at a time, have to be handled, and the frequency of the operation then can be controlled by optimizing or by changing the amount to be handled at one time.

Gilbreth, at the beginning of the century (57), stated "... lifting 90 lb of brick on a packet (sic) to the wall will fatigue a bricklayer less than handling the same number of bricks one or two at a time...." This remark was, of course, based on purely empirical and subjective observations because work physiological instrumentation available during the lifetime of Gilbreth simply did not permit the objective substantiation of such hypotheses. The bricklaying task was repeated by a group of volunteers during the 1950s (57).

In the Gilbrethian example, "lightness" of the task when handling 90 pounds of bricks (18 bricks) one at a time is illusory. Each time a 5-pound brick is handled, the worker must move, bend, erect, rotate, and so on, approximately 100 pounds of body mass. Thus the task load imposed by the handling of the body itself becomes much more severe than the work stress caused by the material being handled. Approximately 1800 pounds of lifting body mass must be maneuvered to shift 90 pounds of brick.

When 90 pounds of brick was handled at one time, according to the Gilbrethian intuitive prescription, the ratio between body mass moved and inanimate material handled was roughly 1:1. As was to be expected, the physiological response to the task, under the circumstances described, was quite moderate and was essentially a function of the rate of productive work, not of unnecessary and physiologically expensive body movements. However when the task load was increased to 120 pounds (i.e., 24 bricks) handled at the same time, the metabolic cost of the job rose out of proportion to the increment in the dynamic component of the task (Table 4). This apparent paradox was rationally explained at the time of the experiment as a result of oxygen debt. The sig-

Table 4 Results of a Replication of the Gilbrethian Lifting Experiment (57)

Frank Bunker Gilbreth, the father of motion study, was intuitively right when he stated that "... to lift 90 pounds of brick at a time is most advantageous physiologically as well as economically ..." (57).

	I	II	III
Bricks per lift	1	18	24
Weight per lift (lb)	5	90	120
Work per hour (kcal)	520	285	450
Bricks per hour	250	600	300

nificance of the element "frequency of lift," as shown previously, can easily cause the handling of numerous small loads to become the equivalent of a very stressful task. It is therefore essential to be aware of the concept of "optimal load" per lift.

5.2 Queueing Situations

In materials handling situations, loads are frequently passed from one worker to the other, or they arrive at the lifting station by way of conveyor belts and are then handled manually. Much useless materials handling takes place when conveyor belts, device pallets, trolleys, and other devices, or areas for temporary holding or storage of products, are too small. Bottlenecks occur, and emergency measures may have to be taken to clear the congested areas. The dimensions of all temporary storage areas and devices should be computed on the basis of queueing theory, assuming that the arrival at the handling station follows the Poisson distribution, and service times are exponentially distributed. Frequently the following formula can be used to advantage (58):

$$N = \frac{\log P}{\log R}$$

where N = required capacity of the area
P = greatest acceptable probability that the area will become temporarily overloaded; this principle is normally determined on the basis of a subjective management decision
R = mean arrival rate of units per time divided by the mean processing rate; these values are normally available from industrial engineers

It should not be forgotten that in addition to readily discernable or "overt" lifting tasks, workplace design may embody hidden or "covert" situations. A covert lifting task exists whenever a moment is exerted on the axial skeleton without an extraneous object being handled. This may be the case when a typist bends her head over the typewriter, or an arm has to be held out for extended periods of time because of the poor location of storage bins, or working shoes have heels that are too high, so that they increase the concavity of the lordotic curve of the lumbar spine (31). To assess the true severity of such a situation, textbooks of classical biomechanics (18) or computerized mathematical models should be consulted (39).

In conclusion, it is reemphasized that the elements of a lifting task are heterogeneous in their physical dimensions and their physiological effects; therefore, although all contribute to the level of work stress, the individual effects cannot be "summed." The reality of most materials handling situations will demand that each element be considered separately and that those which are amenable to control be reduced in magnitude and severity as far as possible.

Unfortunately, at the present state of the art, there are only partial and no total solutions available to eliminate hazards from manual lifting and materials handling tasks.

6 HAND TOOLS

Toolmaking is probably as old as the human race itself. Almost all the basic tools used today, as well as the "basic machines," were invented in the dawn of the prehistoric ages, evolving gradually and parallel with the development of technology until modern times.

However the current technology explosion proceeds too rapidly to permit the gradual development of tools to suit the new industrial processes. Now "instant" creation of new and specialized implements often becomes an acute economic necessity, if an industry wishes to accommodate itself to the rapid changes of workforce and technology. Hand tools are more often than not standardized to be "fairly acceptable" to the broadest possible spectrum of populations and activities. Only rarely are they designed to fit perfectly the specialized needs of a particular manufacturing pursuit or the anthropometric attributes of the workforce in a given plant.

Thus considering their widespread use, the number of varieties of hand tools is quite limited, and each species of tool is normally employed by vast numbers of users. Therefore hand-tool-generated work stress, trauma, and ergogenic disease may at times reach epidemic proportions, disabling numerous individuals, seriously impairing the productive capacity of a manufacturing plant, and having very detrimental effects on labor–management relations. All this is quite apart from the medical costs and the human suffering involved. Therefore the occupational health specialist should make the proper selection, evaluation, and usage of hand tools one of his major concerns.

6.1 Basic Considerations in Tool Evaluation

Though the shapes of tools are varied and the functions of the different classes of implement diverse, there are nevertheless many principles of biomechanics and ergonomics that are applicable to the prevention and solution of problems created by hand tools, no matter how different their fields of application.

Many of the principles enunciated in this section extend in application beyond the narrow field of hand tools proper. They are equally useful in the analysis of equipment controls, since the latter are merely the "tools" that permit the operation of machinery. The initial purpose of primeval tools, such as stone axes and scrapers, was to transmit forces generated within the human body onto inanimate materials, food, or live animals of prey. As the spectrum of artisanal and industrial pursuits widened, the basic purpose of tools became more varied, and today these implements are designed to extend, reinforce, and make more precise range, strength, and effectiveness of limb movement engaged in the performance of a given task.

In this context, the word "extend" does not solely imply a magnification of limb function. Often tools such as tweezers and screwdrivers also make possible far smaller and finer movements than the unarmed hands would be capable of performing. An even better example is the micromanipulator of the "master–slave" type, which serves as attenuator rather than amplifier of human force in motion. A third example is a suction tool that makes it possible to transport small, fragile, and soft workpieces without injury either to their dimensions or their surface finish.

Selection and evaluation of all hand tools, whether manual or power operated, should be based on all the following: technical, anatomical, kinesiological, anthropometric, physiological, and hygienic considerations (59).

In the course of technological development, tool designers were conditioned to focus their attention on a real or imagined need to maximize the tool force output obtainable from a minimal muscular force input from the hand. This, however, should not be overdone.

The operation of a tool should always require sufficient force to provide adequate sensory feedback to the musculoskeletal system in general, and the tactile surfaces of the hand in particular. This is frequently a process of optimization. For example, if a fine screw thread is tapped by hand and the handle of the threading tool is too large, the force acting on the tool becomes excessive, resulting in stripped threads, broken taps, or bruised knuckles. If, on the other hand, the ratio of force output to force input is too small, an unduly large number of work elements must be repeated, and this makes the job fatiguing. An example would be the pounding of a large nail with a very small hammer.

A tool should provide a precise and optimal stress concentration at a specific location on the workpiece. Thus, up to a certain limit, an ax should be as finely honed as possible to fell a tree with a minimum number of strokes, but the edge should not be so keen that it requires frequent resharpening or is fragile. Preferably the tool should be shaped so that it will be automatically guided into a position of optimal advantage where it will do its job best without bruising either hand or workpiece. The Phillips screwdriver, as compared with the ordinary, flat, blade tool, illustrates the latter point.

Hand tool usage causes a variety of stress vectors to act on the man–equipment interface. These may be mechanical, thermal, circulatory, or vibratory, and they are often propagated to quite distant points within the body, far from the actual locus of application of force. The cause of severe pain in the neck muscles or numbness and tingling in the fingers of the left hand may quite conceivably be due to the transmission of somatic resonance vibrations triggered by a vibrating hand tool held in an unergonomic configuration in the right hand. Numerous other examples could be cited. Whenever a single specific anatomical region becomes the locus of repeated manifestations of signs and symptoms of trauma, no matter how far away from the tool operating hand, the work situation should be carefully analyzed for the possible implication of hand tool design or usage as a traumatogenic vector.

Contact surfaces between the tool and the hand should be kept large enough to avoid concentration of high compressive stresses (Figure 45). Pressure and impact acting on the hand may be transmitted either directly or by rheological propagation on vulnerable structures. A poorly designed or improperly held scraping tool may squeeze the ulnar artery and sometimes the ulnar nerve between the timber of the handle and the bones of the wrist. This may deprive the ring and little fingers of proper blood supply (Figure

ERGONOMICS

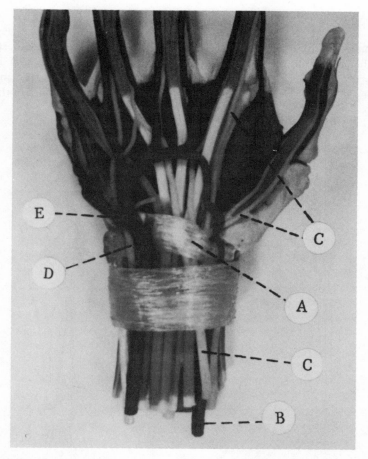

Figure 45 Through the carpal tunnel A pass many vulnerable anatomical structures: blood vessels B and the median nerve C. Outside the tunnel, but vulnerable to pressure, are the ulnar nerve D and the ulnar artery E. From Reference 48.

33), which may cause numbness and tingling of the fingers. Under such circumstances, the afflicted worker will devise an excuse to leave the workplace temporarily, since this is the only means of relief open to him. Apart from the resulting drop in productivity, the health of the working population is in serious jeopardy. Literature (60) suggests that compressive stresses applied against the medical side of the hook of the hamate (Figure 50) can traumatize the ulnar artery and may result in thrombosis or other irreversible injury. The ulnar artery and nerve may also be injured indirectly by stress propagation whenever the palm of the hand is used as a hammer or repeatedly pushes a tool against strong resistance. The resulting lesion, be it vascular or nervous, is known as "hypothenar hammer syndrome" (61).

6.2 The Anatomy of Function of Forearm and Hand

Further chances of injury exist when the motions inventory demanded by specific features in the design of the tool is not readily available from arm and hand. The kinesiology of the upper extremity is basically determined by the structure and arrangement of the skeleton of arm and hand (Figure 46). There are two bones in the forearm,

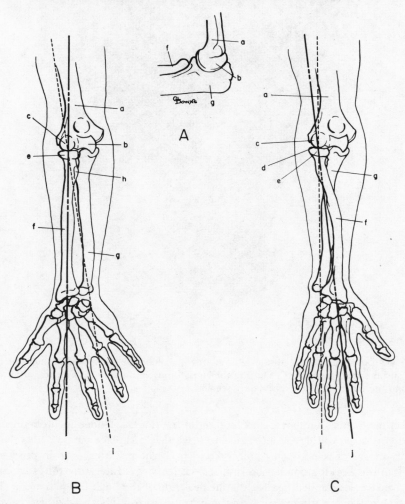

Figure 46 The construction of the skeleton of the forearm. (A) The hinge joint between ulna and humerus from the medial side. (B) The right forearm outwardly rotated. (C) The right forearm medially rotated: a, humerus; b, trochlea of humerus; c, capitulum of humerus; d, thrust bearing formed by capitulum and head of radius; e, head of radius; f, radius; g, ulna; h, attachment of biceps; i, axis of rotation of forearm; j, optimal axis for thrust transmission. From Reference 48.

ERGONOMICS

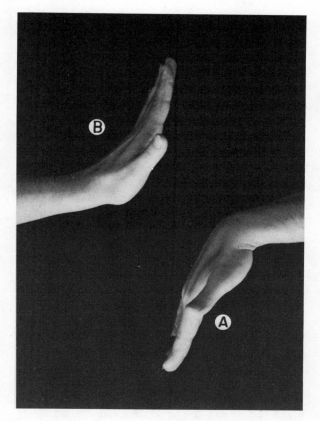

Figure 47 (*A*) Hand in palmar flexion. (*B*) Hand in dorsiflexion.

the ulna and the radius. The ulna is stout at its joint surface of contact with the humerus and forms a hinge bearing there. It is, however, slender at the distal end, where it articulates well with the radius but only poorly with the carpus formed by the bones of the root of the hand. The radius, on the other hand, is slender at its point of contact with the humerus and stout at its distal end. At the proximal end it forms a thrust bearing with the humerus and a journal bearing with the ulna. The distal end forms a joint with the carpal bones, the primary articulation between forearm and hand. The unique configuration of the mating surfaces of this joint permits movement in only two planes, each one at an angle of approximately 90° to each other. The first of these maneuvers is palmar flexion, or when performed in the opposite direction, dorsiflexion (Figure 47). The second set of wrist movements possible consists of either ulnar or radial deviation of the hand (Figure 48). The wrist joint does not allow rotation of the carpus about the longitudinal axis of the forearm. Swiveling the wrist without forearm rotation is not possible (Figure 49).

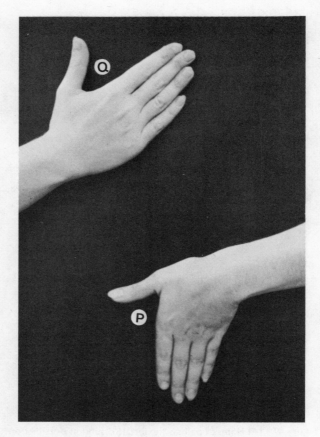

Figure 48 P, ulnar deviation; Q, radial deviation.

This geometry of joint movement causes the axis of longitudinal rotation of the forearm–hand aggregate to run roughly from the lateral side of the elbow joint through a point at the base of the ulnar side of the middle finger (Figure 46). The palmar aspect of the carpal bones forms a concave surface roofed by the transverse carpal ligament. The resulting channel is known as the "carpal tunnel." Through this conduit pass the tendons of the flexor muscles of the fingers, which originate from the medial side of the elbow. Some blood vessels and nerves also pass through this tunnel. A similar, albeit much shallower, passage for the extensor tendons and some nerves is formed on the dorsal surface by another ligament and the dorsal aspects of parts of the carpus, the radius, and the ulna (Figure 50). Each of the two passages is further divided into several longitudinal compartments, and this produces considerable friction and lateral pressures between tendons, tendon sheaths, adjacent nerves, and vascular structures. If the fist is forcefully opened, closed, or rotated, these forces, as well as the friction between ana-

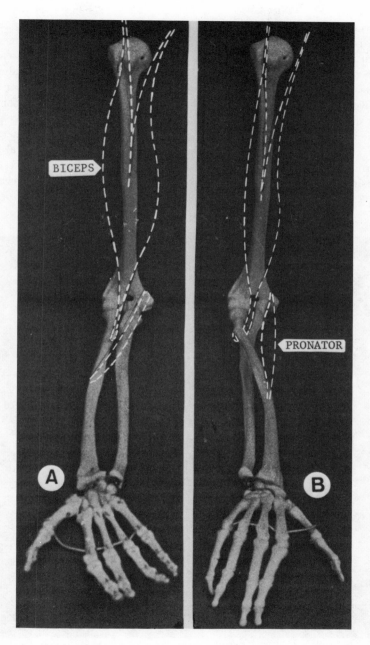

Figure 49 The bones of the wrist articulate with only one of the two long bones of the forearm, the radius, with which they form a firm aggregate. Therefore swiveling the wrist without rotating the forearm is impossible. (*A*) Forearm in supination. (*B*) forearm in pronation.

65

tomical structures, can become unduly high whenever the carpal tunnel and its homologue compartment on the dorsum of the hand are not properly aligned with the longitudinal axis of the forearm. Such alignment exists only if the wrist is kept perfectly straight, so that the metacarpal bone of the ring finger is reasonably parallel with the distal end of the ulna. A potentially pathogenic situation may exist whenever manipulative maneuvers, especially forceful ones that require ulnar deviation, radial deviation, palmar- or dorsiflexion, either singly or in combination with each other, are performed. The misalignments of tendons at the wrist under such circumstances and their bunching up against each other multiply drastically the already high interstructural forces and frictions produced by the muscles operating the hand; early fatigue may result, among other undesirable manifestations (Figure 51). The long interval of time (several weeks, months, even years) of repetitive multiple work stress elapsing before occupational

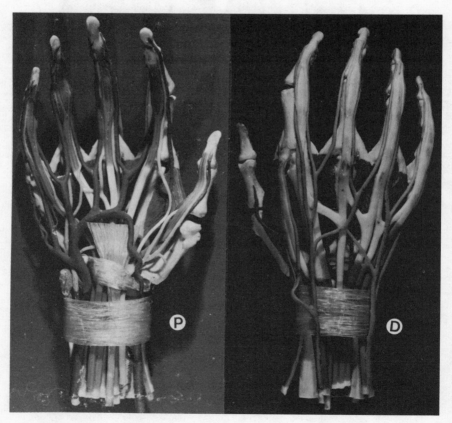

Figure 50 The complex arrangements of tendons, blood vessels, and nerves underneath the ligaments of the wrist: *P*, palmar aspect; *D*, dorsal aspect. Whenever the wrist is deviated, these bunch up against each other. The ensuing friction may lead to trauma and disease.

ERGONOMICS

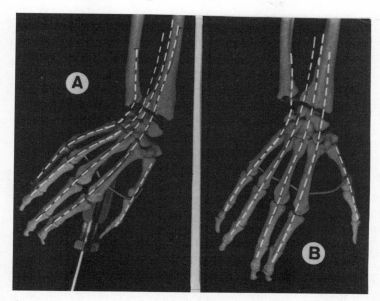

Figure 51 The need to align a tool with the axis of the forearm often forces the hand to deflect toward the ulna (A). The tendons operating the fingers get "kinked" and bunch up. This causes friction between these delicate anatomical structures which, in turn, produces discomfort and, occasionally, diseases of the wrist. When the wrist can be maintained in a straight configuration (B) by good tool design, the tendons are well separated, run straight, and can operate efficiently. From Reference 37, by permission. © American Institute of Industrial Engineers, Inc., Norcross, Ga.

disease of forearm or hand becomes clinically evident, has militated against the definitive establishment of direct cause and effect relationships between the forearm–wrist configuration and specific pathology. Nevertheless, certain motion elements are almost inevitably associated with fairly narrow spectra of occupational disease of the hand (2).

6.3 Elemental Analysis of Hand Movements

The design of tool and handle determines, limits, and defines the motion elements that are necessary for the purpose of the productive process. This intimate relationship between man and hand tools affects occupational health and safety most directly.

It is almost impossible to describe innocuous as well as potentially pathogenic manipulative maneuvers, either verbally or by way of two-dimensional pictorial representation. As a somewhat simplistic yet practical alternative, recourse to the analysis of manipulative maneuvers in terms of their pertinence to "clothes wringing" (Figure 52) is recommended. "Clothes wringing" has been associated for more than a century with tenosynovitis and other undesirable conditions of the hand (2). Here we assume that the wringing is done by a clockwise movement of the right fist and counter-clockwise action of the left.

Figure 52 Any manipulative motion element involving the wrist that may be considered to be part of a "clothes wringing" operation puts the worker at risk.

Whenever a detailed elemental analysis of hand motions is performed, it should be remembered that stress injurious to the hands is produced by four basic conditions:

1. Excessive use against resistance.
2. Hand use while in a potentially pathogenic configuration.
3. Repetitive maneuvers and cumulative work stress rather than "single episode" overexertion.
4. Use of the hand in an unaccustomed manner, as in training, for example (Figure 22.35).

Figure 52 shows that in "clothes wringing" the right hand is engaged simultaneously in supination, ulnar deviation, and palmar flexion. This motion pattern is frequently associated with occurrences of tenosynovitis of the extensor tendons of the wrist or the abductors of the thumb, the latter affliction also known as De Quervain's disease. It is an inflammation of the synovial lining of the tendon sheaths or associated structures. It becomes frequently manifest under conditions of unaccustomed hand usage in new employees (Figure 35) and is therefore occasionally considered quite without justification to be "a training disease." However the basic cause of much exposure to excessive work stress resides not in the learning process but in the misuse of the hand, which has been forced by poor workplace or tool design into unergonomic configuration.

The similarity between the movement of the right hand motion in "clothes wringing" and the use of pliers when looping wires around pegs is quite evident. Since three factors are implicated in the generation of the undesirable work stress, elimination of one of them may successfully reduce the stress to a nonpathogenic level. In this case ulnar deviation was eliminated by bending the tool handle (Figure 35).

ERGONOMICS

In some other work situations involving the same wrist configuration, it may be more desirable, or easier, to reduce excessive palmar flexion by small modifications in workplace layout. A simple change of the distance between worker and work bench, location of the work chair, work surface height, or the degree of tilt of the workspace, may constitute adequate remedial action (40). The experienced practitioner will apply the same principles to the improvement of other pathogenic work situations, such as the manual insertion of screws, the manipulation of rotating switches, or the operation of electrical or air-powered nut setters suspended over the workplace.

In Figure 52 the left hand is engaged in pronation, radial deviation, and dorsiflexion of the wrist. This configuration is conducive to pressures between the head of the radius and the mating joint surface of the humerus (Figure 46). This posture should be strongly discouraged through proper tool design because it may produce a high incidence of the group of diseases known in the vernacular as "tennis elbow." The condition may occur in such tasks as overhead use of wire brushes in maintenance work. This is a good example of the basic principle that work strain may affecct a site distant from the location of work stress. The wrist is being stressed, but it is the elbow that gets sore (62). Strong and repeated dorsiflexion of the wrist, especially in combination with some other hand or forearm movement, is conducive to carpal tunnel syndrome—a disease that may also be provoked by direct trauma to the region of the hand over the carpal tunnel, compression of the median nerve in the tunnel through tenosynovitis and swelling of the flexor tendons, as well as several other causes that are not occupationally related (61). Implicated in unergonomic imposed patterns conducive to carpal tunnel syndrome are tossing motions, the operation of valves located overhead or on vertical walls, and a number of poor handle designs of power tools. This list, however, is by no means exhaustive.

Finally, the wrists of both hands can be severely stressed when a two-handled tool is designed so that the longitudinal axes of both handles coincide. Under such circumstances, the included angle between the axes of the handles should approximate 120°.

6.4 Trigger-Operated Tools

Occasionally the condition of "trigger finger" is encountered. The afflicted person typically can flex the finger but cannot extend it actively. It must be righted passively by external force. When snapped back in such a way, an audible click may be heard. This has been attributed to the generation of a groove in the flexor tendon which snaps into a constriction produced by a fibrous tunnel guiding the tendon along the palmar side of the finger. Small ganglia arising in the tendon sheath have also been implicated (63). This affliction has been observed under conditions of overusage of the index finger as an equipment control, such as the trigger of a tool embodying a pistol grip. The association between overusage of the index finger and the lesion seems to occur most frequently if the tool handle is so large that the distal phalanx of the finger has to be flexed while the middle phalanx must be kept straight. This can easily be the case when females, with relatively small hands, operate tools designed for males. A tool handle a shade too large may put a female working population to distress, but the tool designer who errs on the side of smallness will produce an implement that can be operated with equal effectiveness by both sexes (48). As a rule, frequent use of the index finger should be avoided, and thumb-operated controls should be put into tools and implements wherever possible, because for all practical considerations, the thumb is the only finger that is flexed, abducted, and opposed, in addition to muscles crossing the wrist, by strong, short muscles located entirely within the palm of the hand. It can therefore actuate push buttons and triggers repeatedly, strongly, without fatigue, and without exposure to undue hazard.

6.5 Miscellaneous Considerations

Numerous handheld tools, especially of the power-driven type, are advertised as "light." The lightest tool is not always the best tool. Optimal tool weight is dependent on a number of considerations. If a power tool houses vibrating components, it should be heavy enough to possess adequate inertia. Otherwise it may transmit vibrations of pathogenic frequency onto the body of the operator (see Section 4.3) (42, 43).

Heavy tools should be designed so that the center of mass of the implement is located as close as possible to the body of the person holding it. When this type of facilitation is inadequate, recourse should be taken to suspension mechanisms and counterweights. Strength of handgrip is the most important single factor limiting the weight of a tool that must be held without external assistance from supporting mechanisms. Under such circumstances, a reliable and specialized reference work should be consulted (64).

The design of the tool–hand interface should be based on carefully selected anthropometric considerations. There are a number of specialized anthropometric reference works available (65–67). However not all the hand dimensions identified in literature are necessarily representative of a working population of specialized age, sex, or ethnic origin. Furthermore, the geometry of manipulative movements imposed on the hand by the specifics of tool design or usage may force the user population to employ a geometry of hand movement based on special anthropometric parameters quite different from those available from the majority of reference works. Likewise, the use of working gloves changes the dimensions of the hand. Gloves vary in design and thickness of material used. This does affect grip strength and sensory feedback from the tool (see Section 4.4).

It is recommended that producers as well as consumers of handheld tools conduct their own anthropometric testing based on the hand and other body dimensions of the specialized working population under study, as well as of the tools and gloves to be used. Of course basic reference charts can be employed with advantage as the point of departure for anthropometric hand studies designed to suit specialized needs (68).

Gloves may affect the working hand in a number of additional ways. Mild trauma due to pressure may be aggravated through contacts with irritants entrapped unintentionally in working gloves. In the case of abrasives, such as metal chips or mineral particles, the foreign substances may be worked into the skin, producing in some cases benign tumors, such as talcum granuloma. Certain fat-soluble chemicals, such as many common solvents and detergents, may be soaked up in the material of the glove and transferred onto the skin, causing maceration, dermatoses, or other tissue damage. Sometimes it may be advisable to wear thin cotton gloves under these working gloves to absorb perspiration and improve hygiene.

Any study of work stress, occupational disease, or safety involving hand tools is incomplete unless the potential effects of working gloves are fully considered, inasmuch as they could modify the fit between tool and hand, change the distribution of pressure, or become detrimental to the integrity of the skin.

7 CHAIRS AND SITTING POSTURE

Many jobs require performance in the seated posture, and chairs are among the most important devices used in industry. They determine postural configuration at the workplace as well as basic motion patterns. A well-constructed chair may add as much as 40 productive minutes to the working day of each productive individual (48). Furthermore, poorly designed seating and inadequate policing of seating posture constitute frequent and definitive occupational hazards. Properly designed working chairs are a prerequisite to the maintenance of occupational health and safety of many working populations, including individuals with preexisting disabilities of the back.

A number of reference works (69, 70) provide dimensions for working chairs and benches considered by many as optimal. However such optimality can never be general; it applies only to the restrictive parameters of a population defined by age, sex, ethnic origin, cultural background, and specific working conditions. Sliderulelike devices (71) are available that permit the adaptation of anthropometric data to a wide variety of body dimensions, and in some countries the anthropometric basis for industrial seating has been subject to national standardization (72, 73). However all dimensional aids to chair design must be personalized and revalidated for each individual application on the basis of certain aspects of functional and surface anatomy.

7.1 Anatomical, Anthropometric, and Biomechanical Considerations

It was realized almost a century ago, that ". . . our chairs almost without exception are constructed more for the eye than for the back . . ." (74). This statement was made in an article appearing in a journal of preventive medicine and stating the urgency of providing better lumbar support to seated workers. Unfortunately, in many instances, the same plea can be made today with respect to some of the most modern office and factory furniture. Somewhat later Strasser (75) quantified the forces exerted by the backrest and the seat on the lumbar region and the buttocks. He also showed how these changed drastically as a function of the slope of these features. The first comprehensive study of the biomechanics of seating was not conducted until 1948 (45). This existing body of knowledge should be put to good use in the design of working chairs.

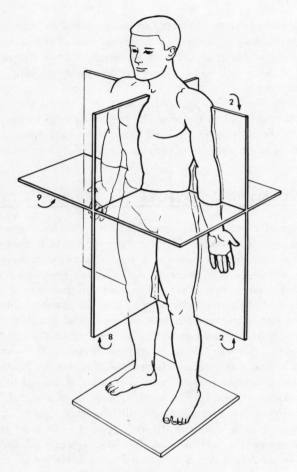

2 = CORONAL PLANE
8 = MID-SAGITTAL PLANE
9 = TRANSVERSE PLANE

Figure 53 The basic planes of reference for biomechanical and anatomical description. From Reference 76, by permission. © W. B. Saunders Company, Philadelphia, Pa.

To facilitate a description of seated work situations, a nomenclature of planes of reference suitable for the definition of relationships between postural configuration and the position of equipment controls, as agreed by convention, is employed throughout this section (Figure 53). The "axis of support" of the seated torso is a line in a coronal plane passing through the projection of the lowest point of the ischial tuberosities on the surface of the seat. This is, in essence, a "two-point support" (Figure 54). As a result, the compressive stresses exerted on the areas of the buttocks underlying the tuberosities is

ERGONOMICS

quite high and has been estimated as 85 to 100 psi. This, of course, varies with body weight and posture. Stress can be nearly double when a person is sitting cross-legged. These high pressures make it necessary to vary sitting posture and position on a seat frequently to provide the necessary stress relief for the body tissues. Therefore the seat of the chair should be approximately 25 percent wider than the total breadth of the buttocks. To further facilitate change of position, all coronal sections of the seating surface of the chair should be straight lines. A coronally contoured seating surface tends to restrict postural freedom at the workplace and, especially when poorly matched to the curvatures of the buttocks, may cause severe discomfort. Improper contouring may also interact with sanitary napkins and other devices worn by women during the menstrual period, with the ensuing further reduction in physical well-being during an already trying time.

The height of the seat has been the subject of much argument, and literature abounds with numerical data, most stating the relevant dimension as the vertical distance of the highest point of the seat from the floor. This is ergonomically wrong. The back of the thighs is ill-equipped to withstand pressure (Figure 55)—compression applied there may deform the limb severely, irritate important nerves and blood vessels, and interfere with the circulation in the lower extremity. Many people had the experience that a chair too high causes the leg "to go to sleep." This is an especially undesirable condition when the circulation is already impaired by preexisting disease, such as diabetes or varicose veins.

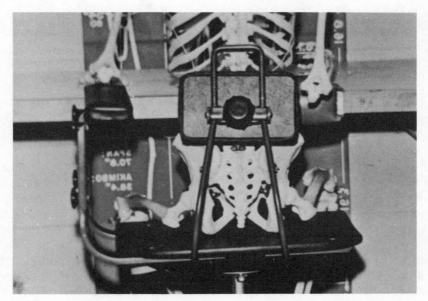

Figure 54 The structure of the pelvis and the location of the ischial tuberosities demand that the coronal sections of a chair be not contoured.

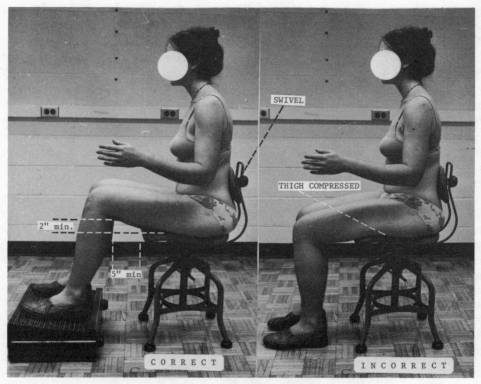

Figure 55 The highest point of the seat should be at least 2 inches below the popliteal crease of the worker. If necessary, this must be accomplished by a footrest. The backrest should swivel about the horizontal axis to align with the lumbar curve.

The maximum elevation of the seat above the surface supporting the feet—be this the floor or a footrest—should be 2 inches less than the crease at the back of the hollow of the knee, known as the popliteal crease. For this reason, the seat height of the chair should be adjustable in a limited number of discrete steps (Figure 55). The frontal end of the seating surface should terminate in a "scroll" edge, which does not cut into the back of the thigh. For further protection, there should be a distance of at least 5 inches between the scroll and the popliteal crease. The seating surface should have a backward slant of approximately 8° in a sagittal direction. This encourages the use of the backrest and prevents forward sliding of the buttocks. Since the thighs are tapered, any inadvertent forward sliding may produce compression of the limb between the lower edge of the work bench and the top of the chair. The seating surface should be slightly padded and covered with a porous, rough, fabric that "breathes" and facilitates adequate conduction of heat away from the contact area between buttocks and chair.

ERGONOMICS

The backrest of a chair employed in manufacturing operations should provide lumbar support. It should be small enough not to exert pressure against the bony structures of the pelvis or the rib cage. At least the top of the backrest should be below all but the "false" ribs. Many work situations require continuous and rhythmic movement of the torso in the sagittal plane, and a backrest that is too high will produce bruises on the backs of a considerable number of the working population.

The best designed backrests are so small that they do not interfere with elbow movement and can swivel freely about a horizontal axis located in the coronal plane (Figure 55). Thus they fit well into the hollow of the lumbar region and provide the needed support for the lower spine without detrimental interference with soft tissues. When much materials handling and twisting of the torso in the seated position takes place a backrest that is overly wide will make frequent and repeated contact with the breasts of female workers (Figure 56). This can be most uncomfortable, especially during the premenstrual period, when the breasts of many women are quite tender.

Whenever backrests produce either bruising or even only slight discomfort, workers protect themselves. The painful effects of excessive interaction between chair and human body are reduced under such circumstances by ad hoc protective devices, such as pillows brought from home and strapped to the backrest. Work sampling studies show that several hours per worker per week may be wasted by the effort needed to keep such improvised cushioning devices in place. Properly designed backrests are much cheaper than improvisation, both in the long run and in the short run.

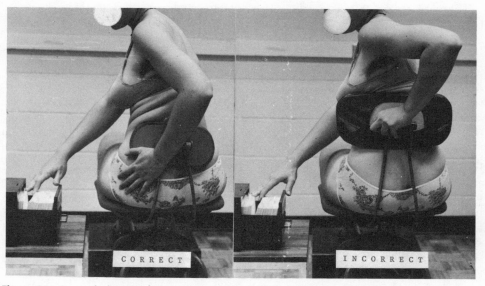

Figure 56 A poorly located backrest that is overly wide may severely traumatize the breasts of female workers during torsional movements. (77)

Some working chairs are equipped with castors. These make possible limited mobility of the worker and permit materials handling without abandoning the seated posture. In many situations castors reduce unnecessary torsional moments acting on the lumbar spine. Also, in some cases of circulatory disturbance in the lower extremity, a chair that can be pushed around by leg movement, while the person maintains seated posture, can activate the "muscle pump," consequently improving circulation. The disadvantage of any chair mounted on rollers resides in the risk that the chair may roll accidentally away or be inadvertently removed while the user gets up for a brief time. If the same person afterward tries to sit down without being aware of the changed situation, a dangerous fall can result. Whenever possible under such circumstances, therefore, a cheap restraining device, such as a nylon rope, or chain or, sometimes, a more rigid linkage between chair and work bench should be considered.

7.2 Adjustment of Chairs on the Shop Floor

A frequent but sometimes dangerous response to chair-generated discomfort is temporary absenteeism from the workplace. When the level of personal tolerance has been exceeded, many individuals simply "take a walk." It is found occasionally that individuals involved in accidents are at the place of injury without authorization. Since no accident is possible unless victim and injury-producing agent meet at the same spot and at the same time, temporary absenteeism from the workplace may result in unnecessary exposure to potentially hazardous situations (40).

To guard against this, especially in new plants or when existing working chairs are being replaced by different models, the workforce should be polled and a subjective assessment of chair comfort established through attitudinal measurement techniques specifically designed for the evaluation of seating accommodations (78). Although subjective comfort evaluations decrease the longer a chair has been in use, it should also be remembered that a comfortable chair is not necessarily the "best chair," regardless of whether high levels of productivity or optimal physiological compatibility with the anatomy of the worker are the guiding considerations. Therefore it is essential that the necessary initial attitudinal measurement be either preceded or immediately followed up by rational adjustment of the seating accommodations as related to the work bench on the basis of biomechanical and ergonomic considerations.

Supervisors as well as workers should receive formal instruction in the proper adjustment of working chairs. First, the height between seating surface and top of the work bench is adjusted so that an angle of abduction of approximately 10° between the upper arms and the torso can be maintained during activity (Section 4.2). In specialized work situations, adjustment for an angle of abduction other than optimal may be necessary. It should be taken into consideration that the natural motions pathway of the wrist changes with the angle of abduction (Figure 57), and the implements of work—components, tools, jigs, and fixtures—should be arranged accordingly on the work bench, otherwise imprecise movements, poor eye–hand coordination, and early fatigue will result. Second, correct seat height with respect to the floor as well as the popliteal

ERGONOMICS

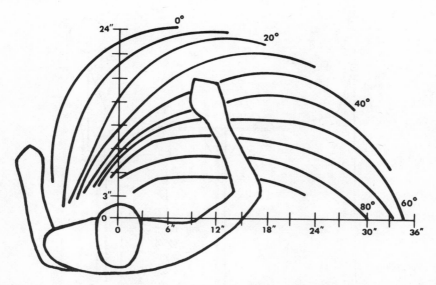

Figure 57 The natural motions pathway of the wrist changes with the angle of abduction. From Reference 48.

crease is established by means of an adjustable footrest. Footrails, because they cannot be easily adjusted, are less desirable. Occasionally drawers located underneath work benches may interfere with proper seat height adjustment, and these should be removed if necessary. Third, the height and position of the backrest should be arranged so that the minimal distance of 5 inches between the front edge of the seat and the popliteal crease is maintained. The adjustment of the backrest should enable the worker to have it low enough to permit freedom of trunk movement as demanded by the work situation, but not so low that it interferes with the bony structures of the pelvis.

Sometimes it is desirable to provide the opportunity to perform a job from either the seated or standing posture, or to alternate both postures in the course of the working day, according to the personal preference of the worker. Then the basic angle of abduction of the upper arm should be attained by a suitable height of the work bench. To maintain this for both postures, the chair and the footrest should be of the appropriate dimensions (Figure 58).

Finally, the distance of the backrest of the chair from the near edge of the work bench should be standardized to optimize skeletal configuration (Sections 4.2 and 4.4), to facilitate the development of the most effective kinesiology.

7.3 Ancillary Considerations

No single chair design can possibly be optimal for all work situations. Seating analysis should always be conducted to meet the needs of specific, not generalized situations, giv-

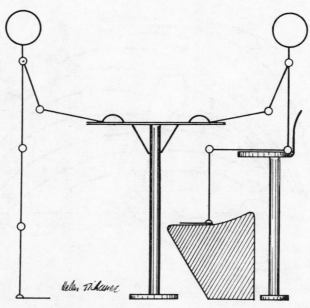

Figure 58 When work is possible in either seated or standing position, work bench and seating design should permit change of posture without change of musculoskeletal configuration. From Reference 48.

ing due weight to all relevant features of the task under consideration. It is highly desirable that standard reference works (68–70) be available for perusal.

In addition to the analysis of the seated posture with respect to physical comfort and biomechanical correctness, it is necessary to consider the changes in kinesiology of the lower extremity resulting from seated posture. The seated leg and foot can rotate only with difficulty, and unless the height of the chair is extremely low, such as is the case in motor vehicles, operation of foot pedals may become cumbersome and fatiguing. On the other hand, the seated "leg and thigh aggregate" can abduct and adduct voluntarily, precisely, and strongly, without fatigue for long intervals of time. It is therefore frequently advantageous to make use of this kind of movement in the design of machine controls. In the seated posture, knee switches, such as are often employed in the operation of industrial sewing machines, are generally superior to foot pedals.

8 ERGONOMIC EVALUATION OF WORK SITUATIONS

Historically, all industrial and technological development in the United States has been triggered by the need to overcome and solve problems in the design, production, distribution, and use of manufactured articles. Traditionally, industry has subscribed to the "improvement approach" as the principal avenue toward economic efficiency and viability of American enterprise.

Before the development of industrial and human engineering, it was common practice to conceive products hastily, determine an adequate manufacturing process intuitively, and gradually remove deficiencies in product design or manufacturing methods during actual production. Such improvement extended over a period of several months or even years. This approach was to some extent acceptable in past decades, and the model T Ford remained in production for approximately 30 years. In today's fast-changing marketplace, this kind of policy often leads to economic disaster. By the time a product has been improved, it may have become redundant. Unfortunately the "improvement" approach to product design and manufacture is still maintained in many industries under the guise of the term "continuous cost reduction." This often conveniently excuses failure to design the product and process correctly before production.

In some industries "cost reduction" is expected to be practiced routinely by supervisory and engineering personnel during the first months or years of production of a new article. This tempts those who are responsible for "efficiency" to design new products and work methods initially imperfect, so that easy opportunities for later cost reduction are not lost.

Unfortunately, a similar attitude often prevails when the occupational health and safety of the working population is at stake. Too often industry waits for evidence of work-induced occupational disability before commissioning an occupational health specialist to identify the causes, which are then removed by a process of not always satisfactory gradualism.

Only too often the practitioner of ergonomics is challenged to justify the removal of pathogenic vectors from the working environment by a prediction of potential savings in medical costs, or an increase of productivity likely to accrue from his activities. It is therefore important that those who are engaged in the practice of ergonomics be ready to prove that the maintenance of occupational health and high levels of productivity are inseparable. Three main areas of ergonomic evaluation are of prime interest to the practitioner in industry: (*a*) historical evaluation, (*b*) analytical evaluation, and (*c*) projective evaluation.

8.1 Historical Evaluation

An active interest of management in ergonomics is often initially triggered by noticeable breakdowns in occupational health occurring in the performance of a well-defined, essential, and easily identifiable productive operation. This then results in increased manufacturing expense, high medical cost, potential retribution by a regulatory agency, and other undesirable side effects. Under such circumstances, a request for historical evaluation of past activities and events may be made. Such study is best conducted keeping in mind the "four big C's" of occupational health investigation:

1. Cause.
2. Consequence.
3. Cost.
4. Cure.

Frequently consequence and cost are known, and the cause and the cure remain to be discovered. Here theoretical analysis is the most expedient tool of research. Experimental methods are generally an unnecessary expense and quite superfluous in developing a critique of past events. Furthermore, a theoretical analysis offers a degree of confidentiality not available in experimentation with man.

It must be reemphasized that this book concentrates only on those narrower aspects of ergonomics (79) that are related to its subspecialty: biomechanics (80). To treat all the disciplines tributary to ergonomics and their applications within the confines of the space available would have led to a superficial discussion or sometimes mere mention of numerous important aspects of the industrial environment, its evaluation, and control.

It is important that no evaluation of work situations be undertaken unless the investigator is familiar with the ergonomic aspects of climate control, noise, vibration, illumination, circadian rhythms, and related topics. All these should be included in an environmental workup, which should always precede the general study of physical interaction between man and the implements of the workplace. Figure 2 should provide helpful guidance in the systematic conduct of such a preliminary workup. The next step would be to ascertain whether the prerequisites of biomechanical work tolerance (Section 4.1) were to some extent disregarded in the design of the product manufactured or the work method. A step-by-step checkout of each situation with Table 2 in hand is the best approach; afterward it should be possible to suggest some potential causes for the observed anatomical, physiological, or behavioral failure. Sometimes it may then be possible to suggest remedial action immediately. However the use of tables and "cookbooks" is no substitute for speculative analysis based on sound professional knowledge.

Superficial study of occupational accident or disease vectors based on mere guidelines may very often obfuscate cause and effect relationships. What may appear to be the cause of an accident could be, in fact, an effect produced by a less obvious mechanism. Discussion of an actual case may illustrate the need to probe energetically for the primary cause of occupational injury.

In a chemical factory the number of workers hit by forklift trucks while crossing aisles increased suddenly and dramatically without any apparent cause. Whenever such a "lost time" accident report was filed, either pedestrian error or driver error was listed as the cause. Subsequently, common human factors engineering approaches likely to eliminate the problem were explored. The trucks were made more visible by painting them in conspicuous colors, and illumination in the aisles was improved. At some of the places of greatest accident frequency, automatic warning horns were installed which signaled whenever a vehicle approached. When none of these measures proved to be successful, investigators began to try to discover why so many individuals were walking

ERGONOMICS 81

around the factory instead of remaining seated safely at their workplaces. Accident frequency was proportional to the number of pedestrians present in the aisles at any given time. Brief periods of absenteeism from the workplace suddenly increased dramatically, leading in turn to an increase in pedestrian traffic density.

The time of this change coincided with the introduction of a new tool (see Figure 33). An electrical brush used to clean trays was replaced by a much less expensive but equally effective paint scraper that produced insults to the ulnar artery (Section 4.3). This reduced blood supply to the ring and little fingers. The resulting numbness and tingling caused the individuals afflicted to lay down their tools occasionally and seek relief by exercising their hands. To avoid ensuing arguments with supervisors, workers were tempted to make use of every opportunity of brief absences from the job. Trips to the washroom, the toolroom, and so on, became much more frequent, and this was the true cause of increased exposure of the factory population to the risk of traffic accidents. Thus the first of the four big C's was identified. The cure: the handle of the paint scraper was redesigned. The result: the workers spent more time per day in productive activity; thus the output and economy of the operation increased, while at the same time, because of diminished risk exposure, the accident rate returned to normal.

Whenever the frequency of incidence of occupational ill health or accident increases after a manufacturing process has been in safe operation for some time, the following question should be asked: *what change in equipment, product design, tools used, working population employed, or work method applied has taken place immediately before the breakdown of occupational health?*

8.2 Analytical Evaluation

The procedures of analytical evaluation are called for whenever an existing manufacturing operation is generally satisfactory but has to be improved to make it more competitive, to reduce training time, or to eliminate operator discomfort and ill health.

Theoretical analysis (Section 8.1) is the initial step in all analytical evaluation. However this should often be followed by some additional procedures. The simplest and perhaps most effective aid in this kind of study is cinematography and subsequent frame-by-frame analysis. This permits a detailed evaluation of the workers' reaction to each event at the workplace and to each contact with tool, machine, or manufactured article. Slow-motion viewing of the operation not only reveals biomechanical or ergonomic defects but is instrumental in discovering reflex reactions to mild localized repetitive trauma that cannot be detected by the naked eye because of the brief duration of many such events. Stage magicians know that the hand is quicker than the eye. The current trend toward the use of video tape for work analysis should be discouraged because the tapes are inadequate for the detection of fine details of expression (81) or blanching of the skin, or frame-by-frame analysis. Furthermore, color video taping is exorbitantly expensive, whereas color film, especially in the Super-8 size, is economical and tells much more than a black and white picture. Finally, manufacture of a video tape from movie film is very inexpensive, whereas the converse (i.e., the manufacture of a movie film from video tape) is a very costly operation. Furthermore, being magnetic, video tape requires more careful storage, is sensitive to magnetic fields, and often is erased accidentally.

It is important for motion picture analysis of a work situation to allow the viewing of the workplace in at least two different planes, if necessary, with the aid of mirrors. When motion picture analysis alone is not adequate for process evaluation, recourse to other experimental technologies must be taken. Some investigative procedures, such as metabolic measurement, electromyography, and electromyographic kinesiology, have already been described (Section 3). This work can often be complemented by the production of biomechanical profiles, which are of special usefulness when the potential pathogenic effects of individual and brief motion elements of work are under discussion (37); therefore their main field of application resides in projective evaluation.

8.3 Projective Evaluation

Whenever possible, a job should be ergonomically evaluated while it is still in the planning phase. This makes it possible to "design out" of a task features, equipment, and maneuvers that are potentially traumatogenic. Projective evaluation should include reliable predictions with respect to the work tolerance of a specific population, estimated duration of training, and counseling procedures useful in overcoming difficulties in the training process.

All projective evaluation should include, when necessary, biomechanical profiles (82, 83), both as a means to establish reliable effort input–production work output relationships and to predict potential anatomical failure points.

The profile is a polygraphic recording produced during the performance of a standard element of work that includes displacement, velocity, and acceleration of at least one anatomical reference point. Such measurement is performed by kinesiometers—apparatus that permits the performance of a specific motion element against a known resistance. *Displacement* is indicative of range and pattern of motion; *velocity* serves as an index of speed as well as strength (slow joint movement is often associated with muscular weakness) (37).

Finally, *acceleration* reflects control over precision and quality of motion (84). Abnormal acceleration and deceleration signatures are invariably associated with imprecise and unsafe movements due to the inability to terminate a motion at the correct place and time. These biomechanical parameters are recorded simultaneously with integrated myograms of selected muscles involved in the performance of the task under study. The usefulness of the biomechanical profile becomes evident when a tossing motion is investigated. Many individuals have difficulty with wrist extension, but this does not interfere with their competence to perform assembly tasks in an entirely satisfactory manner. Once the task is finished, however, the individual may not be able to dispose of the article by tossing it into a bin (Figure 59). Inability to toss is more frequent than is commonly assumed. If caused by common industrial disorders, such as

ERGONOMICS

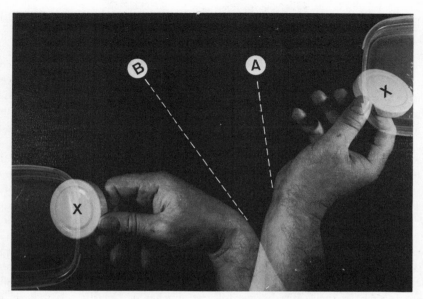

Figure 59 Hyperextension of the wrist is not a recognized element in motion study, but it is often essential for disposal of items at workplaces X. Hand and wrist disease are frequent occurrences, preventing the successful execution of this biomechanical motion, especially in women and aged workers. A simple change of location of the disposal bin may make the difference between occupational disability A or ability B. From Reference 37, by permission. © American Institute of Industrial Engineers, Inc., Norcross, Ga.

tenosynovitis, it will persist throughout the working day. Alternatively, when caused by fatigue, it may become evident only after several hours of work have elapsed.

To establish a cause and effect relationship between the muscular effort involved in wrist extension and the pattern and quality of the tossing motion produced, a biomechanical profile is constructed. The kinesiometer employed for this purpose appears in Figure 60. The subjects are tested before the start of the working day, then retested after several hours of normal productive work. The manner in which the task affects the performing individual can be established by comparison of the "prework" and "postwork" biomechanical profiles (Figure 61).

In the case of a tendency toward or a history of tenosynovitis, characteristic changes in the profile can be observed with ease. Displacement shows no change. However, the myograms are slightly stronger, indicative of a greater effort necessary to produce movement. The velocity curve displays notches at peak speed. This demonstrates that multiple, repetitive efforts must be made during each movement to achieve peak performance. The biomechanical profile cannot be used to make a definitive diagnosis such as tenosynovitis. The results obtained permit only the statement that some anatomicomechanical obstruction interferes with the tossing motion. This could be caused by a large number of conditions—orthopedic, arthritic, or others. Medical diagnosis, of course, is under the jurisdiction of a physician.

Fatigue changes the profile in quite a different fashion. No meaningful change whatsoever can be observed in the type of myogram used here. However the displacement tracing shows that it is not possible to maintain wrist extension for the necessary interval of time, and the acceleration signature displays evidence of muscular rigidity and fine tremors (Figure 61). This establishes that the task is fatiguing. Either the rhythm of the work cycle should be changed or the length of work periods and rest pauses should be significantly modified. A less expensive, less complex, and more feasible way to deal with both classes of disability may be a simple change in the position of the disposal bin, which would simply eliminate the necessity for wrist extension at the end of the work cycle (Figure 59). This demonstrates that the biomechanical profile, as distinct from other methods of performance measurement, not only indicates that quality and/or magnitude of performance are defective, but also makes it possible to pinpoint the physical cause of the deficiency so that successful remedial action may be taken.

The biomechanical profile, one of the newest procedures in ergonomic work measurement, bridges the gap that has existed since the beginning of scientific management, industrial psychology, and work physiology, plaguing most practitioners in industry. Workers as early as the Gilbreths had fully established a scientific rationale on which the disciplines and biomechanics as practiced today are based. Nevertheless, they lacked

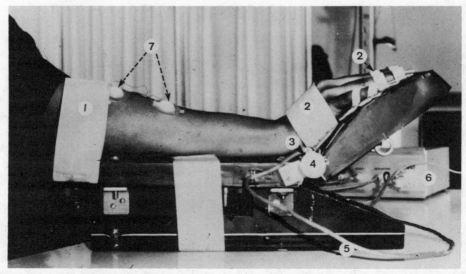

Figure 60 Wrist extension kinesiometer: 1, arm stabilizer; 2, wrist and finger stabilizers; 3, hinge under radiocarpal joint producing 10 lb-in.; 4, electrogoniometer; 5, cable to computation module; 6, analogue computation module generating biomechanical profile; 7, electrodes for extensor myogram. From Reference 84.

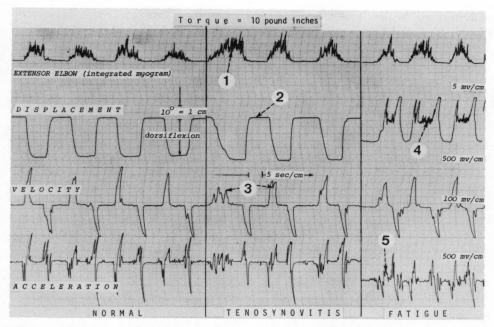

Figure 61 Biomechanical profile produced by the wrist extension kinesiometer in Figure 60: 1, increased extensor myogram; 2, normal displacement signature; 3, notched velocity peaks; 4, inability to maintain wrist extension; 5 signs of rigidity in acceleration signature. From Reference 84.

instrumentation adequate to conduct experimental investigations into the physical effort expended by individual muscles in the performance of a specific task. The analytical thinking of these pioneers was simply 50 years ahead of the technology available at that time. The second industrial revolution has introduced productive processes that constrain the worker to a relatively rigid posture, which forces him to maintain repetitive motion patterns throughout a long working day. This has produced or aggravated numerous known, as well as previously unknown, elements and complaints. However the new technologies (85, 86), as a by-product, have produced instrumentation now available for effective ergonomic work measurement. These make possible the prevention of occupational disability, increasing levels of physiological and emotional well-being of the working population, as well as the productive capacity and the competitive posture of American enterprise.

ACKNOWLEDGMENTS

This book is based on three main references (37, 38, and 48) which were developed from the lecture notes I prepared for students at the University of Queensland, the University of New South Wales, Texas Tech University, and New York University. Most of the graphic work and illustrations were contributed by my wife Helen. Special thanks are due to Audrey Lane for the painstaking editorial efforts taken with this manuscript.

GLOSSARY

The purpose of this glossary is to bridge the communication gap between the technological and health professions. Although all the terms included can be found in general or medical dictionaries, the context in which they are used changes the meanings of many items of scientific jargon. Furthermore, an encyclopedic explanation provides, in many instances, information excessive to such a degree as to obfuscate those aspects important to a reader with specialized interests. The definitions in this glossary are intended to satisfy the specialized needs of the practitioner in fields related to the health, well-being, and productivity of man at work.

ABDUCTION: Antonym of adduction.
ACHILLES TENDON: Connects the principal plantar flexor muscles of the foot acting on the ankle joint with the heel of the skeleton of the foot.
ACTION POTENTIAL: Response to a stimulus applied to a nerve or muscle cell. Commonly expressed in millivolts and defined by the resulting changes in potential difference between the inside and the outside of the cell, related to a time base. Sometimes its maximal value is incorrectly named "action potential."
ACTIVITY TOLERANCE: The ability to perform a task effectively without undue strain and at acceptable levels of emotional and physiological well-being.
ADDUCTION: Movement that reduces the angle between a limb and a sagittal plane (q.v.).
ALL OR NONE LAW: Whenever a stimulus weaker than threshold level is applied to excitable tissues, including nerves and muscles, there will be no response at all. However, a stimulus above threshold level will, irrespective of its strength, cause a complete physiological reaction. For example, a stimulated muscle fiber will either not contract at all or completely contract.
ALVEOLUS: In the context of this text, terminal blind sac of a bronchiolus, the finest branch of the windpipe.
ANATOMY OF FUNCTION: Description of the operating characteristics of anatomical structures in response to the demand of the physical environment.
ANTAGONIST: Muscle that opposes a movement caused by protagonist (q.v.).
ANTHROPOMETRY: The body of knowledge pertaining to human body measurements and sometimes to the weight of body segments.
ARTHRITIS: In its broadest sense the term means "inflammation of a joint." Within this book the term is used for the group of degenerative joint diseases usually associated

with progressing age. *Note:* It should not be confused with rheumatoid and other forms of arthritis.

ATLANTOOCCIPITAL JOINT: Joint formed between the first cervical vertebra and the base of the skull, facilitating head movement in the sagittal plane (q.v.).

AXIAL SKELETON: The skull without the mandible, the vertebral column and the sacrum being considered as a single structural unit.

AXON: The long process of a nerve cell. Conducts impulse.

BEHAVIORAL REACTION: Changes in behavior such as precision of movement consistently associated with specific features within the working environment or task cycles.

BICEPS: Long twin-bellied muscle going from the shoulder blade to the proximal end of the radius, thus crossing and acting on both the shoulder and the elbow joints. Often mistakenly assumed to be the principal flexor of the forearm, which it is not. (See BRACHIALIS). The biceps is a strong supinator of the forearm, rotating the flexed forearm outward.

BIOELECTRIC: Refers to electric activity of living tissues, for example, electrocardiogram, electroencephalogram, and electromyogram.

BIOMECHANICAL LIFTING EQUIVALENT: Magnitude of moment exerted on the lumbar spine by a load, irrespective of the weight of the object. Expressed in pound-inches or corresponding metric units.

BRACHIALIS MUSCLE: Robust short muscle. Origin covers lower third of anterior surface of humerus. Inserts into anterior part of ulna close to elbow joint. Strong principal flexor of forearm operating the hinge joint formed between ulna and humerus.

BURSA: Small bag filled with fluid, reducing friction between moving structures.

BURSITIS: Inflammation of a bursa. Common occupational disease. Diverse types of bursitis in the vernacular include tennis elbow, housemaid's knee, tailor's bottom.

CAPILLARY: Very small blood vessel that connects the smallest branches of the arteries with those of the veins.

CAPITULUM OF HUMERUS: Spherical prominence located on the anterior aspect of the distal end of the humerus. Forms thrust bearing with the head of the radius.

CARPAL TUNNEL: Channel on the palmar side of the wrist formed by the irregular small bones of the wrist and a tough ligament stretched across it. Through the carpal tunnel pass the flexor tendons of the fingers, the median nerve, and some blood vessels.

CARPUS: The aggregate of eight small irregular bones forming the wrist.

CIRCUMDUCTION: Movement about a ball and socket joint, which causes a limb to move so that its distal end describes a circle.

CONCENTRIC CONTRACTION: Increase of tension within a muscle producing shortening. For example, the brachialis shortens when a weight is lifted by flexing the forearm.

DELTOID: Large muscle of the shoulder that abducts and otherwise moves the upper arm about the shoulder joint against gravity.

DERMATOSIS: Disease of the skin.

DE QUERVAIN'S DISEASE: Tenosynovitis (q.v.) of the abductors of the thumb. Frequently incorrectly used as a synonym to tenosynovitis in general.

DISABILITY: Incapacity to perform some or all of the sensory and/or motor functions necessary for occupational pursuit or successful function in activities of daily living. Frequently but not necessarily the result of impairment (q.v.).

DISTAL: In a limb: further away from the body. Elsewhere: further away from the central axis of the body.

DYNAMIC WORK: "Work" according to the definition in mechanics. Defined as the product of a force multiplied by the distance through which its point of application moves. Units of work commonly used in ergonomics and biomechanics: calorie, joule, foot-pound, metre-kilogram.

ELECTROGONIOMETER: Device to quantify in analog fashion or digitally an angle and changes of angle between body segments connected by a joint. Frequently a potentiometer is the transducer employed.

ELECTROMYOGRAM: Electrical potential produced during contraction of tissue, consisting of skeletal muscle and made visible by galvanometers or oscilloscopes. Depending on signal condition, useful in the diagnosis of neuromuscular disease or in the determination of strength and sequencing of contraction of a number of muscles under investigation.

ELECTROMYOGRAPHIC WORK MEASUREMENT: Loosely used term. No real measurement of work performed takes place, but the order of magnitude and sequencing of muscular activity are made visible in analog form through the integrated "electromyogram" (q.v.).

ELECTROMYOGRAPHY: The recording of action potentials (q.v.) from contracting muscles for the purposes of medical diagnosis for the analysis of motion patterns.

ELECTROPHYSIOLOGICAL KINESIOLOGY: Formerly known as "electromyographic kinesiology": the study of kinesiological phenomena through the analysis of strength, frequency, and magnitude of electromyographic and other electrophysiological data.

END ORGAN: Ending of a terminal branch of a sensory nerve specialized so that it can sense only one specific type of stimulus; for example, those end organs in the skin that perceive heat cannot feel cold.

EPICONDYLITIS: Technical term for "tennis elbow."

ERGOGENIC DISEASE: Disease produced by exposure to a work situation in which ecological stress vectors cause disease.

ERGONOMICS: A multidisciplinary activity dealing with the interactions between man and his total working environment, plus such traditional and environmental aspects as atmosphere, heat, light, and sun, as well as of tools and equipment of the workplace. (From American National Standard ANSI Z794.1-1972.)

EXCENTRIC CONTRACTION: Increase of tension within a muscle while lengthening. For example, the branchialis exerts a force resisting the pull of gravity, lowering a weight slowly while the flexed forearm extends.

EXCITATION: In the context of electromyography, the act of stimulation of a nerve or muscle cell resulting in the generation of bioelectricity.

FATIGUE: Imprecise term, loosely used, but often implying an involuntary decrement of response to a social stimulus (e.g., incentive payment), a biological stimulus, or imprecise as well as unpredictable response to environmental demand.

GANGLION (pl. GANGLIA): A cyst in a tendon sheath or joint capsule.

GONIOMETER: Device measuring the angle and range of angular movement between two body segments connected by a joint.

HAMATE: Small bone on the ulnar side of the wrist. Has a small hook-shaped process called the "hook of the hamate" on its palmar side. The hook shields the ulnar artery and ulnar nerve against pressure.

HEAD OF HUMERUS: That end of the humerus (q.v.) which forms a joint with the shoulder blade.

HEAD OF RADIUS: Essentially disk shaped, proximal end of radius, held in place by a circular ligament, acts as a rotating shaft articulating with the proximal end of the ulna. The proximal surface of the head is slightly concave and, when the angle between the forearm and upper arm is approximately 90°, it forms a thrust bearing with the capitulum of the humerus (q.v.).

HUMAN PERFORMANCE: In industrial usage an expression of output from physical work in terms of economic efficiency.

HUMERUS: Long bone of the upper arm.

IMPAIRMENT: Deficit of physiological function. Distinct from and often unrelated to disability (q.v.).

INDUSTRIAL PSYCHOLOGY: The application of the body of knowledge of psychology to specific problems observed in industrial settings. The term is frequently used loosely and imprecisely. Sometimes it is employed interchangeably with human factors engineering. Occasionally it carries a connotation of the study of perceptive and cognitive problems or a special emphasis on sensory motor performance.

INTERPHALANGEAL JOINTS: Joints connecting two knuckles.

ISCHIAL TUBEROSITIES: Two bony prominences forming the lowest point of the pelvis. On them rests the weight of the body when seated, and important muscles are attached to them.

ISOMETRIC WORK: Term often misunderstood by engineers. Describes the condition when a muscle exerts a force (i.e., contracts) against resistance without producing any motion, for example, to hold a dumbbell still with the extended arm. Isometric work, which results in increased demand for calories, is different from work in mechanics, defined as force multiplied by the time of its application and measured in units such as kilogram-seconds.

JOULE: Unit of work; 4000 joules (abbreviated J), are approximately equal to 1 kilocalorie.

KINESIOLOGY: The study of human movements as a function of the construction of the musculoskeletal system.

KYPHOSIS: Pronounced convexity of the spine. Most frequently observed in the thoracic region. Also known as "hunchback."

LIGAMENTS: These frequently connect one bone with another bone and may either facilitate or limit movement.

LOCOMOTION: The act of moving the animal body from place to place using the musculoskeletal system.

LORDOSIS: Concave curvature of the spine. Exists in the neck and in the lumbar region. Conversationally the term is frequently used to indicate excessive lumbar lordosis.

LUMBOSACRAL JOINT: Joint between fifth lumbar vertebra and sacrum.

MEDIAL: Reference to that side of an anatomical structure which is closest to the midsagittal plane.

MEDIAN NERVE: Large important nerve. Activates muscles that pronate the forearm and flex forearm, wrist, and fingers. The sensory part of the nerve provides feedback information from the thumb and the first two and one-half fingers essential for prompt and effective performance of the "grasp reflex."

METABOLISM: Biomechanical process(es) that converts foodstuff into tissue elements, or is involved in the physiological "combustion" producing energy available for work.

MOTOR CELL: Nerve cell especially adapted for the conveying of excitatory impulses to muscle. (See EXCITATION).

MOTOR UNIT: The body of a nerve cell plus its axon (q.v.), plus all muscle fibers supplied by branches of one axon, that is, the functional "building block" of muscle.

MYOELECTRICITY: Loosely used term. May refer either to action potentials during excitation or a single muscle fiber, or to the generation and changes in level of electric voltages produced during contraction of a mass of muscle.

MYOGRAPHY: Term frequently used as synonym of electromyography (q.v.).

NOX (pl. NOXAE): Any agent injurious to safety or health.

NERVE ROOT ENTRAPMENT SYNDROME: Technical term for "pinched nerve."

OCCUPATIONAL DISEASE: Disease caused or precipitated by exposure to the occupational environment, equipment, or motion patterns necessary for the performance of a task. The difference between occupational accident and disease is not a difference in kind; the word "accident" merely indicates that the interval of time for the development of the pathological condition is very short. A hand gradually crippled over the years by traumatic arthritis is just as disabled as one suddenly smashed by a hammer.

PALPATE: To locate by touch.

PATHOGENIC: Producing disease.

PATHOLOGY: The discipline dealing with the development and description of disease in terms of altered structure and function of the body. Sometimes used loosely to denote a diseased state.

PECTORALIS MAJOR: Large triangular muscle. The base forms the origin, running parallel to the entire length of the breast bone. The apex inserts into the medial side of the humerus. Essentially an adductor of the upper arm.

PHALANX (pl. PHALANGES): Small bones of the fingers and toes. In this book the two most distal bones of the thumb and the three most distal bones of the other four fingers.

POSTPRANDIAL: Pertaining to the condition of the stomach and its contents as well as to the general physiological status of an individual, as related to the time elapsed since the last meal.

PHYSIOLOGICAL RESPONSE: Adaptation of a physiological system to work stress, for example, constriction of the blood vessels of the hand during exposure to cold or vibrations.

PHYSIOLOGICAL SYSTEM: The aggregate of organs, tissues, and anatomical structures in general involved in a specific function, essential to the normal process of living. An example is the nervous system, the system integrating and cordinating activities of the animal body.

PHYSIOLOGY: The study of chemical and physical behavior of anatomical structures and body systems under conditions incidental to the process of living.

POPLITEAL: Pertaining to the hollow at the back of the knee.

PRONATION: The action of rotating the flexed forearm toward the midsagittal plane, so that the hands become prone, with palms down, back of hand up.

PRONATOR-TERRES: Spindle-shaped muscle running from the medial side of the humerus above the elbow to the lateral side of the radius in its middle. Produces medial wrist rotation, that is, pronation.

PROTAGONIST: Muscle initiating and producing a motion.

PROXIMAL: In a limb, closer to the body. Elsewhere, closer to the central axis of the body.

PULMONARY: Pertaining to the lung.

PUMONARY VENTILATION: Synonym for respiratory minute volume. The amount of air inhaled per minute, usually expressed in liters. Not to be confused with tidal volume or vital capacity.

RADIUS: One of the long bones of the forearm. The slender proximal end forms a thrust bearing with the end of the humerus and a rotating bearing between a circular ligament and the proximal end of the ulna. The broad distal end forms a joint with the irregular small bones, which form the root of the hand and another joint with the distal end of the ulna.

SAGITTAL PLANE: Anatomical reference plane vertically dividing the body into right and left portions.

SCAPULA: Shoulder blade.

SCOLIOSIS: Lateral curvature of the spine. A frequent anatomical abnormality, when concurrent with kyphosis (q.v.) it produces hunchback. May impair respiratory function.

SESAMOID BONES: Lentil-shaped bones embedded in tendons. Reduce friction between a tendon and joint surfaces during friction. For example, the patella is the large sesamoid bone of the knee joint.

SOMA: Body.

SOMATIC: Pertaining to the body.

STANDARD ELEMENT OF WORK: Repetitive sequence of basic motions or sensory activities which form part of a work cycle. Standard elements of work have been identified in various systems of industrial work measurement.

SUPINATION: Process of rotating the flexed forearm outward so that hand becomes "supine," that is, "palms up." During supination the radius swivels around the ulna.

SYNOVIA: Membranes lining the inside of joint capsules and moving surfaces of joints. They secrete the synovial fluid, which lubricates joints.

TENDON: Band of connective tissue that does not contract and connects muscles with bones.

TENDON SHEATHS: Tubular structures through which tendons run. They are lined with a synovial membrane and, therefore, not only guide but also lubricate the tendons.

TENOSYNOVITIS: Inflammation of the tendon sheaths.

TENSION-TIME: Concept accepted by physiologists as representative of the level of isometric work performed by a muscle or a muscle group. Equals tension in units of force multiplied by the duration of the contraction measured in seconds. Thus tension-time is the area under the strength–duration curve.

THROMBOSIS: The narrowing of a blood vessel caused by the formation of a clot.

TRANSDUCER: Device employed in instrumentation capable of converting an input signal into visible display (such as a pressure gauge) or changing its nature (e.g., a thermocouple converts heat into an electric current).

TRAUMA: In this book mechanical injury produced by any physical agent.

TRAUMATOGENIC: The property of producing trauma.

TRICEPS: Three-headed large extensor muscle of the forearm. Originates from the back of the humerus and the shoulder blade. Inserts into the proximal tip of the ulna and opens the hinge joint formed between the forearm and the upper arm.

ULNA: Long bone of the forearm. Stout at proximal end and forms a fairly rigid hinge joint with the humerus, which is only involved in flexion and extension of the forearm. The distal end is slender and forms a joint with the radius but not with the bones of the hand.

VECTOR: Used in the life sciences with a connotation different from that in engineering. Any external agent exerting an influence on body function (from the Latin "carrier"). For example, in tropical medicine the mosquito is the transmitting vector of malaria).

VISCUS (pl. VISCERA): Generic name of the large organs located within the four body cavities: the skull, the thorax, the abdomen, and the pelvis.

WORK PHYSIOLOGY: Application of the body of knowledge of physiology to the study of physiological phenomena commonly associated with "work." Often used loosely and as a synonym of exercise physiology, sometimes falsely carrying a restrictive connotation implying nothing more than measurement of metabolic and associated phenomena during physical activity.

WORK STRAIN: Physiological response to work stress, for example, increased heart rate when climbing stairs. Work strain always results from the performance of physical work, and is undesirable only when it exceeds activity tolerance (q.v.).

WORK STRESS: The application of external force resulting from work situations. Always present whenever a task is performed, and undesirable only when excessive.

WORK TOLERANCE: Synonym of activity tolerance (q.v.) when applied to work situations.

REFERENCES

1. B. Ramazzini, *Essai sur les Maladies de Artisans* (translated from the Latin text *De Morbis Artificum* by M. de Fourcroy), Chapters 1 and 52, 1777.
2. D. Hunter, *The Diseases of Occupations,* 4th ed., Little, Brown, Boston, 1969, p. 120.
3. J. Amar, *Organization Physiologique du Travail,* H. Dunod, Paris, 1917.
4. E. D. Adrian, "Interpretation of the Electromyogram", *Lancet,* **2,** 1229–1233; 1283–1286 (1925).
5. O. G. Edholm and K. F. H. Murrell, *The Ergonomics Research Society, A History,* 1949–1970, Wykeham Press, Winchester, 1973.
6. P. V. Karpovich and G. P. Karpovich, "Electrogoniometer: A New Device for the Study of Joints in Action," *Fed. Proc.,* **18,** 311 (1959).
7. A. J. S. Lundervold, *Electromyographic Investigations of Position and Manner of Working in Typewriting,* A. W. Broggers Boktrykkeri A/S, Oslo, 1951.
8. E. R. Tichauer, "Electromyographic Kinesiology in the Analysis of Work Situations and Hand Tools", *Proceedings of the First International Conference of Electromyographic Kinesiology, Electromyography,* Supplement 1 to Vol. 8, 1968, pp. 197–212.
9. M. Lukiesh and F. K. Moss, "The New Science of Seeing," in: *Interpreting the Science of Seeing into Lighting Practice,* Vol. 1, General Electric Company, Cleveland, 1932, pp. 1927–1932.
10. H. S. Belding and T. F. Hatch, "Index for Evaluating Heat Stress in Terms of Resulting Physiological Stress," *Heat. Pip. Air Cond.,* **27,** 129 (1955).
11. B. L. Welch and A. S. Welch, Ed., *Physiological Effects of Noise,* Plenum Press, New York, 1970.
12. W. F. Floyd and A. T. Welford, Eds., *Symposium on Fatigue,* H. K. Lewis & Co., London, 1953.

13. E. R. Tichauer, "Potential of Biomechanics for Solving Specific Hazard Problems," in: *Proceedings 1968 Professional Conference,* American Society of Safety Engineers, Park Ridge, Ill., 1968, pp. 149–187.
14. P. J. Rasch and R. K. Burke, *Kinesiology and Applied Anatomy,* 3rd ed., Lea & Febiger, Philadelphia, 1967.
15. S. Brunnstrom, *Clinical Kinesiology,* 3rd ed., revised by R. Dickinson, F. A. Davis, Philadelphia, 1972.
16. D. L. Kelley, *Kinesiology: Fundamentals of Motion Description,* Prentice-Hall, Englewood Cliffs, N.J., 1971.
17. E. R. Tichauer, *Occupational Biomechanics (The Anatomical Basis of Workplace Design),* Rehabilitation Monograph No. 51, Institute of Rehabilitation Medicine, New York University Medical Center, New York, 1975.
18. M. Williams and H. R. Lisner, *Biomechanics of Human Motion,* Saunders, Philadelphia, 1962.
19. R. Fick, *Anatomie und Mechanik der Gelenke,* Vol. 3, Spezielle Gelenk und Muskelmechanik Fisher, Jena, 1911, pp. 318–389.
20. N. Recklinghausen, *Gliedermechanik und Lähmungsprothesen,* J. Springer, Berlin, 1920.
21. H. A. Haxton, "Absolute Muscle Force in Ankle Flexors of Man," *J. Physiol.,* **103,** 267–273 (1944).
22. R. W. Ramsey and S. F. Street, "Isometric Length-Tension Diagram of Isolated Skeletal Muscle Fibers in Frog," *J. Cell. Comp. Physiol.,* **15,** 11–34 (1940).
23. B. A. Houssay, *Human Physiology,* translated by J. T. Lewis and O. T. Lewis, McGraw-Hill, New York, 1955, p. 385.
24. L. Brouha, *Physiology in Industry,* 2nd ed., Pergamon Press, Oxford, 1960.
25. J. B. de V. Weir, "New Methods for Calculating Metabolic Rate with Special Reference to Protein Metabolism," *J. Physiol.,* **109,** 1–9 (1949).
26. C. F. Consolazio, R. E. Johnson, and L. J. Pecora, *Physiological Measurements of Metabolic Functions in Man,* McGraw-Hill, New York, 1963.
27. F. H. Norris, Jr., *The EMG: A Guide and Atlas for Practical Electromyography,* Grune & Stratton, New York, 1963.
28. J. V. Basmajian, *Muscles Alive,* 3rd ed., Williams & Wilkins, Baltimore, Md., 1974, p. 7.
29. H. H. Ju, "A Statistical Multi-Variable Approach to the Measurement of Performance Effectiveness of a Lumped System of Human Muscle," Master's thesis, New York University, New York, 1970.
30. J. F. Davis, *Manual of Surface Electromyography,* Wright Air Development Center Technical Report No. 59-184, Wright-Patterson Air Force Base, Ohio, 1959.
31. E. R. Tichauer, *Biomechanics of Lifting,* Report No. RD-3130-MPO-69, prepared for Social and Rehabilitation Service, U.S. Department of Health, Education and Welfare, Washington, D.C., 1970.
32. H. O. Kendall et al., *Muscles: Testing and Function,* 2nd ed., Williams & Wilkins, Baltimore, Md., 1971.
33. A. V. Hill and J. V. Howarth, "The Reversal of Chemical Reactions of Contracting Muscle During an Applied Stretch," *Proc. Roy. Soc.,* S.B. 151:169 (1959).
34. B. Jonsson, "Electromyographic Kinesiology, Aims and Fields of Use," in: *New Developments in Electromyographic and Clinical Neurophysiology,* Vol. 1, J. E. Desmedt, Ed., Karger, Basel, 1973, pp. 498–501.
35. B. Jonsson and M. Bagberg, "The Effect of Different Working Heights on the Deltoid Muscle," *Scand. J. Rehab. Med.,* Suppl. 3, 26–32 (1974).
36. E. Asmussen et al., *Quantitative Evaluation of the Activity of the Back Muscles in Lifting,* Communication No. 21, Danish National Association for Infantile Paralysis, Hellerup, 1965.
37. E. R. Tichauer, "Biomechanics Sustains Occupational Safety and Health," *Ind. Eng.,* February 1976.

38. E. R. Tichauer, "Occupational Biomechanics and the Development of Work Tolerance," in: *Biomechanics V-A*, P. V. Komi, Ed., University Park Press, Baltimore, Md., 1976, pp. 493–505.
39. D. B. Chaffin and W. H. Baker, "A Biomechanical Model for Analysis of Symmetric Sagittal Plane Lifting," *AIIE Trans.*, **2**: 1, 16–27 (1970).
40. E. R. Tichauer, "Ergonomics: The State of the Art," *Am. Ind. Hyg. Assoc. J.*, **28**, 105–116 (1967).
41. J. Hasan, "Biomedical Aspects of Low Frequency Vibration," *Work-Environment-Health*, **6**: 1, 19–45 (1970).
42. G. Loriga, in *Occupation and Health, Encyclopedia of Hygiene, Pathology and Social Welfare*, ILO, Geneva, 1934.
43. A. Hamilton, J. P. Leake, et al., Bureau of Labor Statistics Bulletin No. 236, Department of Labor, Washington, D.C., 1918.
44. D. E. Wasserman and D. W. Badger, *Vibration and the Worker's Health and Safety*, Technical Report No. 77, National Institute for Occupational Safety and Health, Government Printing Office, Washington, D.C., 1973.
45. B. Akerblom, *Standing and Sitting Posture*, A.-B. Nordiska Bokhandelns, Stockholm, 1948.
46. E. R. Tichauer, "Some Aspects of Stress on Forearm and Hand in Industry," *J. Occup. Med.*, **8**: 2, 63–71 (1966).
47. B. W. Niebel, *Motion and Time Study*, 4th ed., Irwin, Homewood, Ill., 1967, p. 169.
48. E. R. Tichauer, in: *The Industrial Environment—Its Evaluation and Control*, National Institute for Occupational Safety and Health, Department of Health, Education and Welfare, Washington, D.C., 1973, pp. 138–139.
49. M. Arnold, *Reconstructive Anatomy*, Saunders, Philadelphia, 1968, p. 391.
50. W. T. Dempster, "The Anthropometry of Body Action," in: *Dynamic Anthropometry*, R. W. Miner, Ed., *Ann. NY Acad. Sci.*, **63**: 4, 559–585 (1955).
51. E. R. Tichauer et al., *The Biomechanics of Lifting and Materials Handling*, Report No. HSM 99-72-13, submitted to the National Institute of Occupational Safety and Health, New York, 1974.
52. L. E. Abt, "Anthropometric Data in the Design of Anthropometric Test Dummies," in: *Dynamic Anthropometry*, R. W. Miner, Ed., *Ann. NY Acad. Sci.*, **63**: 4, 433–636 (1955).
53. J. B. Martin and D. B. Chaffin, "Biomechanical Computerized Simulation of Human Strength in Sagittal-Plane Activities," *AIIE Trans.*, 19–28 (1972).
54. I. Starr, "Units for the Expression of Both Static and Dynamic Work in Similar Terms, and Their Application to Weight-Lifting Experiments," *J. Appl. Physiol.*, **4**: 21 (1951).
55. P. V. Karpovich, *Physiology of Muscular Activity*, Saunders, Philadelphia, 1959.
56. I. J. Schorr, *Changes in Myoelectric Activity of the Erector Spinae, Gluteus Maximus and Hamstring Muscles During Pick-Up and Release of Loads for Various Workplace Geometrics*, Master's thesis, New York University, New York, 1974.
57. F. B. Gilbreth, *Motion Study*, Van Nostrand, New York, 1911.
58. E. R. Tichauer, "Industrial Engineering in the Rehabilitation of the Handicapped," *J. Ind. Eng.*, **19**: 2, 96–104 (1968).
59. R. Drillis, D. Schneck, and H. Gage, "The Theory of Striking Tools," *Hum. Factors*, **5**: 5 (October 1963).
60. J. M. Little and A. F. Grant, "Hypothenar Hammer Syndrome," *Med. J. Aust.*, **1**, 49–53 (1972).
61. D. Briggs, "Trauma," in: *Occupational Medicine*, C. Zenz, Ed., Year Book Medical Publishers, Chicago, 1975, pp. 254 ff.
62. E. Grandjean, *Fitting the Task to the Man*, Taylor and Francis, London, 1969.
63. H. Bailey et al., *A Short Practice of Surgery*, H. K. Lewis & Co., London, 1956.

64. F. Fitzhugh, *Gripstrength Performance in Dynamic Tasks,* Technical Report, University of Michigan, Ann Arbor, 1973.
65. C. E. Clauser et al., *Anthropometry of Air Force Women,* AMRL-TR-70-5, Aerospace Medical Research Laboratory, Wright-Patterson Air Force Base, Ohio, 1972.
66. J. W. Garrett, "The Adult Human Hand: Some Anthropometric and Biomechanical Considerations," *Hum. Factors,* **13:** 2 (1971).
67. J. W. Garrett et al., *A Collation of Anthropometry,* AMRL-TR-68-1, 2, 2 vols., Wright Air Development Center, Wright-Patterson Air Force Base, Ohio, 1971.
68. H. Dreyfuss, *The Measure of Man,* 2nd ed., Whitney Library of Design, New York, 1967.
69. K. H. E. Kroemer, *Seating in Plant and Office,* AMRL-TR-71-52, Aerospace Medical Research Laboratory, Wright-Patterson Air Force Base, Ohio, 1971.
70. E. Grandjean, *Fitting the Task to the Man—An Ergonomic Approach,* Taylor and Francis, London, 1967.
71. N. Diffrient et al., *Humanscale 1/2/3,* Henry Dreyfuss Associates, MIT Press, Cambridge, Mass., 1974.
72. *Specification for Office Desks, Tables and Seating,* B. S. No. 3893, British Standards Institution, 1965.
73. *Anthropometric Recommendations for Dimensions of Non-Adjustable Office Chairs, Desks and Tables,* B. S. No. 3079, British Standards Institution, 1959.
74. F. Staffel, "Zur Hygiene des Sitzens," *Z. Allg. Gesundheitspflege,* **3,** 403–421 (1884).
75. H. Strasser, *Lehrbuch der Muskel- und Gesundheitspflege,* Vol. 2, J. Springer, Berlin, 1913.
76. S. W. Jacob and C. A. Francone, *Structure and Function in Man,* Saunders, Philadelphia, 1970, p. 8.
77. S. Slesin, "Biomechanics," *Ind. Design,* **18:** 3, 36–41 (1971).
78. B. Shackel, K. D. Chidsey, and Pat Shipley, "The Assessment of Chair Comfort," in: *Sitting Posture,* E. Grandjean, Ed., Taylor and Francis, London, pp. 155–192.
79. *The Origin of Ergonomics,* Ergonomics Research Society, Echo Press, Loughborough, England, 1964.
80. E. R. Tichauer, in: *Biomechanics Monograph,* E. F. Byars, R. Contini, and V. L. Roberts, Eds., American Society of Mechanical Engineers, New York, 1967, p. 155.
81. R. J. Nagoe and V. H. Sears, *Dental Prosthetics,* Mosby, St. Louis, 1958.
82. E. R. Tichauer, H. Gage, and L. B. Harrison, "The Use of Biomechanical Profiles in Objective Work Measurement," *J. Ind. Eng.,* **4,** 20–27 (1972).
83. E. R. Tichauer et al., "Clinical Application of the Biomechanical Profile of Pronation and Supination," *Bull. NY Acad. Med.,* 2nd ser., **50:** 4, 480–495 (1974).
84. E. R. Tichauer, in: *Rehabilitation After Central Nervous System Trauma,* H. Bostrom, T. Larsson, and M. Ljungstedt, Eds., Nordiska Bokhandelns Förlag, Stockholm, 1974.
85. E. R. Tichauer, M. Miller, and I. M. Nathan, "Lordosimetry: A New Technique for the Measurement of Postural Response to Materials Handling," *Am. Ind. Hyg. Assoc. J.,* **34,** 1–12 (1973).
86. C. Sparger, *Anatomy and Ballet,* A. & C. Black, London, 1960.

Index

For descriptive information of terms not listed in this index, refer to Glossary, pages: 86-93.

Anatomy, of function, 1, 4, 5
    functional, 4
    lever systems, classification, 5, 11
        first-class, 5
        second-class, 5, 6
        systematic, 4
        third-class, 7
        topographic, 4
        torsional, 7

Biomechanical profile, diagnostic, 82
    fatigue, 84
    generation, 82
    kinesiometer, 82
    myogram, 82
    prognostic, 82
        anatomical failure, 82
    repetitive trauma, 85
    signature, acceleration, 82
        displacement, 82
        velocity, 82
    tenosynovitis, 83
Biomechanics, experimental, 2

Chairs, absenteeism, caused by, 76
    adjustment of, 76
    anatomy and, 71
    anthropometry, 71
    axis of support, 72
    backrest, 75
    biomechanics and, 71
    buttocks, stresses on, 72
    castors, 76
    footrail, 77
    footrest, 77
    height, 73, 77, 78
    knee switches, 78
    pedals, operation, 78
    seat, contour, 73
        edge, frontal, 74
        height, 73
        impaired circulation caused by, 73
        slant, 74
        width, 73

Diagram, force-velocity, 11
    length-tension, 10, 11

Electrogoniometer, 2
Electromyogram, analysis, 1
    direct, 27, 28
    evaluation, 28, 30
    integrated, 28, 29, 30
    interpretation, 1, 27, 30
    needle, 21, 22
    quality, 28
    surface, 22, 29, 30
Electromyography, action potential, 20
    apparatus, 22
    electrodes, 21
    measurement of effort, 16
    membrane potential, 20
    motor unit, 20
    muscle testing, 24
    myoelectricity, 20

technique, 22
work measurement, 19
Equipment controls, placement, 11
Ergonomic analysis, 5
  cinematography in, 81
Ergonomic evaluation, analytical, 79, 81
  historical, 79
  projective, 79, 82
  video tape, by, 81
Ergonomics, 2
Ergonomic stress vectors, 3
Ergonomist, 4

Forearm-hand carpal tunnel, 64, 69
  configuration, 64
  joints, 63
  kinesiology, 62
  pathogenic, 66
  radius, 63
  skeleton, 62
  ulna, 63
  wrist movement/swivel, 63

Ganglion, 69

Hand, carpal tunnel, 64, 66
  carpal tunnel syndrome, 69
  carpus, 63
  clothes wringing, 67, 68
  DeQuervain's disease, 68
  deviation, radial, 63, 69
    ulnar, 63, 68
  dorsiflexion, 63, 69
  palmar flexion, 63
  pronation, 69
  supination, 68
  tenosynovitis, 67

Joints, action, 9
  degrees of freedom of motion, 9
  elbow, 10
  force of separation, 14

hip, 10
humeroulnar, 10
movement, safe range, 14
thrust/torque, acting on, 13, 14

Kinematic element, kinematics, 9
Kinesiology, electrophysiological, 2
  force diagram, 13
  lever systems, analysis, 12
  occupational, 9
  workplace layout, 10
Kinetic element, 9, 10

Lifting, aggregation, 56
  body segments, 48
  components, gravitational, 55
  dynamic, 56
  elemental analysis, 47, 48
  equivalent, biomechanical, 34
  forces, inertial, 56
    segregation, 56
  frequency, 56
  isometric, 55, 56
  models, computerized, 54
  moments, lateral, 53
    saggital, 48
    static, 48
    torsional, 53
  negative, 56
  seated, 54
  sex differences, 50
  stress, 34
  work, dynamic, 56
    isometric, 55
    negative, 56

Mechanics, kinematics, kinetics, and statics, 9
Metabolism, bioenergetics, 18
  heavy work, 48
  measurement, 16, 17
  oxygen consumption, 17
  pulmonary ventilation, 18

INDEX

Weir's equation, 18
Movement, isometric, 11
Muscle, function, 14
  insufficiency, active/passive, 14
  strength, 14
  tension, internal, 14

Physiology, definition, 16
  measurement in, 16
  work, 1
Psychology, industrial, 1

Queueing, facilities, capacity of, 58
  theory, 68

Tennis elbow, 69
Tools, contact surfaces, 60
  evaluation, 59, 60
  finger-grooved, 41
  forces, input/output, 60
  purpose, 59
  sensory feedback, 60

stress, due to, 60
traumatizing, 61
trigger-operated, 69
vibrating, 60

Work tolerance, blood flow, 37
  blood vessels, pressure on, 37
  chair design, 41
  fatigue, antagonist, 46
  forward reaches, 43
  gloves, 45
  head movement, 36
  ischemia, compression, 37
  muscular insufficiency, 44
  postular correlates, 33
  prerequisites of, 33
  scanning, visual, 36
  sex differences, 35
  skeletal configuration, 36
  vibration, 39
  work strain/stress, 31. 32